Alert and Ready

An Organizational Design Assessment of Marine Corps Intelligence

Christopher Paul, Harry J. Thie, Katharine Watkins Webb,

Stephanie Young, Colin P. Clarke, Susan G. Straus,

Joya Laha, Christine Osowski, Chad C. Serena

Prepared for the United States Marine Corps
Approved for public release; distribution unlimited

RAND NATIONAL DEFENSE RESEARCH INSTITUTE

The research described in this report was prepared for the United States Marine Corps. The research was conducted within the RAND National Defense Research Institute, a federally funded research and development center sponsored by the Office of the Secretary of Defense, the Joint Staff, the Unified Combatant Commands, the Navy, the Marine Corps, the defense agencies, and the defense Intelligence Community under Contract W74V8H-06-C-0002.

Library of Congress Control Number: 2011928032

ISBN: 978-0-8330-5260-5

The RAND Corporation is a nonprofit institution that helps improve policy and decisionmaking through research and analysis. RAND's publications do not necessarily reflect the opinions of its research clients and sponsors.

RAND® is a registered trademark.

Cover photo by Cpl. R. Logan Kyle, USMC

Published 2011 by the RAND Corporation
1776 Main Street, P.O. Box 2138, Santa Monica, CA 90407-2138
1200 South Hayes Street, Arlington, VA 22202-5050
4570 Fifth Avenue, Suite 600, Pittsburgh, PA 15213-2665
RAND URL: http://www.rand.org/
To order RAND documents or to obtain additional information, contact
Distribution Services: Telephone: (310) 451-7002;
Fax: (310) 451-6915; Email: order@rand.org

Preface

Since 2001, the U.S. Marine Corps (USMC) intelligence enterprise has demonstrated its agility in tailoring its organization to meet evolving expeditionary force demands. This has resulted in a number of ad hoc arrangements, practices, and organizations. Moreover, as the USMC has grown in strength over recent years, it has also added intelligence personnel. The USMC Director of Intelligence asked the RAND National Defense Research Institute to broadly review the organizational design of the USMC intelligence enterprise. The study addressed how to align the organization of USMC intelligence to efficiently and effectively carry out current and future missions and functions. The study was designed to focus on organizational structure and, because of the short duration, to be fairly general in nature. Specifically, it considered the organization of (and possible improvements to) the Intelligence Department, the Marine Corps Intelligence Activity, the intelligence organizations within the Marine Expeditionary Forces (specifically, the intelligence and radio battalions), and intelligence structures in the combat elements.

This research was sponsored by the USMC and conducted within the Intelligence Policy Center of the RAND National Defense Research Institute, a federally funded research and development center sponsored by the Office of the Secretary of Defense, the Joint Staff, the Unified Combatant Commands, the Navy, the Marine Corps, the defense agencies, and the defense Intelligence Community. The principal investigator is Harry Thie. Comments are welcome and may be sent to Harry_Thie@rand.org.

For more information on the RAND Intelligence Policy Center, see http://www.rand.org/nsrd/ndri/centers/intel.html or contact the director (contact information is provided on the web page).

Contents

Figures

Tables

Summary

Background

U.S. Marine Corps (USMC) intelligence personnel collect, analyze, and disseminate intelligence to support USMC operational components and leaders. The geopolitical landscape within which this occurs has changed drastically since the 1994 Intelligence Plan (Van Riper Plan) sought to restructure USMC intelligence in response to perceived shortcomings exposed by the first Gulf War.[1] Today, international security concerns abound, and issues such as the rise of lethal nonstate actors, nuclear proliferation by rogue nations, and shifting power dynamics in strategically vital regions all threaten global stability. These external developments have unfolded alongside an ongoing internal reorganization of the U.S. Intelligence Community (IC), as well as the workforce and structure of USMC intelligence more specifically.

Not only have the threats changed since the implementation of the Van Riper Plan, but the tools needed to counter a diverse array of adversaries have changed as well. Globalization, sophisticated satellite technology, and the ubiquitous reach of the Internet, among other developments, have spawned advances in real-time communication. To meet the demands of this complex security and information environment, the USMC has grown to 202,000 marines, and the number of marines with intelligence military occupational specialties has more

---

[1] See C4I Staff, Headquarters, U.S. Marine Corps, "The Future of Marine Corps Intelligence," *Marine Corps Gazette*, Vol. 78, No. 4, April 1995, pp. 26–29.

than doubled since 1994. Continuous counterinsurgency operations have changed tactical support structures, and technological innovations have provided new tools and capabilities. Furthermore, the USMC has been tasked with taking the lead on issues of cultural intelligence within the broader IC.

With Operation Iraqi Freedom coming to a close and an Operation Enduring Freedom drawdown a distinct possibility, a new USMC force posture will begin to take shape. Despite the possibility that the service will have both less money and fewer troops, the USMC intelligence enterprise will no doubt be called upon to remain alert and ready while "doing more with less," a common theme expressed in interviews and a mainstay of USMC culture.

Purpose of This Research

The USMC asked the RAND National Defense Research Institute to review the organizational design and assess how the USMC intelligence enterprise can more efficiently and effectively carry out current and future missions and functions. The study was designed to focus explicitly on organizational structure. The research considered four organizational levels, depicted in Figure S.1: (1) the Intelligence Department (Director of Intelligence [DIRINT] and immediate staff), (2) the Marine Corps Intelligence Activity, (3) the intelligence and radio battalions, and (4) the combat elements, primarily the ground combat element.

Our findings are based on a review of the literature on organizations and organizational theory, interviews with more than 100 marines or USMC civilians, and a structured assessment process.

Key Findings

The Marine Corps Intelligence Department Reflects an Accumulation of 20 Years of Organizational Change

The USMC Intelligence Department (I-Dept), by virtue of its headquarters placement, focuses more on inputs (e.g., money, manpower)

Figure S.1
Structure of Marine Corps Intelligence and the Four Organizational Levels Analyzed

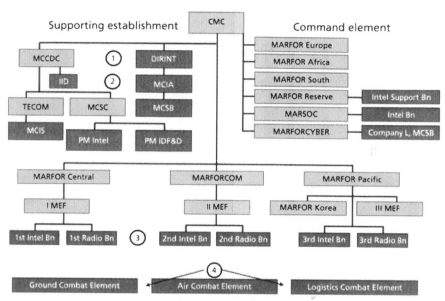

NOTE: Bn = battalion. IID = USMC Intelligence Integration Division. MARFOR = USMC Forces. MARFORCOM = USMC Forces Command. MARSOC = USMC Forces Special Operations Command. MCCDC = USMC Combat Development Command. MCIA = USMC Intelligence Activity. MCIS = USMC Intelligence School. MCSB = Marine Cryptologic Support Battalion. MCSC = USMC Systems Command. MEF = Marine Expeditionary Force. PM IDF&D = program manager, intelligence data fusion and dissemination. PM Intel = program manager, intelligence systems. TECOM = USMC Training and Education Command.
RAND MG1108-S.1

than customers (e.g., the operating forces). Moreover, the I-Dept has grown rapidly and reactively rather than in a planned manner. As a result, names of subunits do not reflect their actual functions, and the organization is somewhat opaque to outsiders, making it difficult to engage. There has been an inconsistent long-term strategic focus on overall IC goals because the various I-Dept offices are more consumed with day-to-day activities.

The Marine Corps Intelligence Activity Lacks Customer Orientation and Has Unclear Priorities

Serving multiple masters complicates coordination processes in MCIA, and resources do not always align with priorities. Its multiple customers (e.g., IC, DIRINT and I-Dept, operating forces) and its functional organization lead to frequent "reach in" by knowledgeable personnel to gain needed data, information, or assistance, to the detriment of overall organizational functioning. Customer service is lacking, and MCIA has neither an effectively oriented web presence nor 24/7 service. Products and services lack functional integration focused on customer needs.

The Focus of the Marine Expeditionary Force is "Up" and Disciplinary

Support of the combat elements is generally described as lacking in that it is not relevant and not timely. Moreover, products are not sufficiently integrated across functions. When there are competing demands, servicing the "up" customer takes priority, irrespective of real need. The intelligence battalion trains as an intelligence battalion but does not deploy as a battalion, while the radio battalion is perceived as residing in its own cocoon.

Combat Elements Have Shifted from a Functional to a Matrix Structure but Are Hampered by a Lack of Experience

Over the past few years, the intelligence structure at the Ground Combat Element (GCE) has shifted from functional to matrix, from a battalion-level functional S2 intelligence structure to a company-level intelligence cell in which intelligence personnel from the battalion S2 section are "matrixed" with infantry marines at the company level. The liability of a matrix structure in the present environment is that intelligence personnel assigned to battalion level need to be experienced and expert in their craft, and that is not always the case.

There Are Other Issues Related to Mission, Workforce, Leadership, Culture, and Technology

Some of these issues might be construed as "organizational" in a broader sense; others, not. We discuss these points throughout this monograph because they have the potential to affect USMC intelligence strategic objectives and thus may require attention or resolution through organizational changes or other approaches. Organizational change could improve performance in these issue areas, or it could be counterproductive and hamper the effectiveness of the organizational changes analyzed in Chapter Seven.

Recommendations

The Intelligence Department Is a Functional Hierarchy and Should Stay That Way, but Opportunistic Improvements Are Needed

The issues and concerns that we identified in I-Dept can be addressed without changing the nature of the department's functional structure, but rather by realigning it. Specifically, several of the resourcing functions could be grouped together. Appropriate roles and reporting relationships should be established for senior civilians. One subunit with an operational orientation (the Intelligence Estimates Branch) could be placed elsewhere because it is functionally different from all other subunits. However, because it supports high-level offices (primarily, the Commandant of the Marine Corps), it is best kept in the I-Dept.

The Marine Corps Intelligence Activity Should Reorganize into a Specialized Matrix Known as a Front-Back Organization

For MCIA, we recommend a structural alternative that is a specialized matrix form called a front-back organization. This structure is designed to accommodate both customer and product effectiveness and functional efficiency. It can also better accommodate absences for training or deployment. Furthermore, it has the advantage of maintaining easy access and habituation with customers but allocates expertise more efficiently, and it allows more functional training and development of expertise because experts are a pooled resource. The ability to manage and monitor customer needs and demands, and to efficiently allocate expertise and resources to meet those demands, is particularly important to MCIA, with its host of varied customers.

Marine Expeditionary Force Could Be More Effective if Organized into Integrated Matrix Habitual Relationships

A significant change at the MEF level would be to integrate functions in the battalion by creating discipline-integrated, company-level units and to associate these units habitually in both general and direct support relationships with particular regimental combat units.[2] In practice, the USMC is familiar with such an integrated structure because it is used elsewhere and is the basic structural form for Marine Expeditionary Unit intelligence capabilities. This structure better supports decentralized decisionmaking and, because the USMC operating concept focuses on the Marine Expeditionary Brigade as the key organization, it provides dedicated and habitual support for that commander.

---

[2] A model for such habituation between formations in the USMC is artillery. However, there are differences in the traditional relationship between artillery and the regimental combat units and what we are proposing for intelligence. Artillery units are one level above intelligence units; we are proposing an intelligence company to support a regimental combat team where, for artillery, it would be a whole battalion. Moreover, artillery has a well-developed doctrine for this support, with fire support coordinators or artillery liaison teams allocated to all levels of the supported organization. Intelligence doctrine only discusses the role of the intelligence battalion commander as the overall intelligence support coordinator for the MEF. (This has its own problems.) If the organizational structure is changed at this level, intelligence doctrine needs to be extended beyond that for the intelligence support coordinator.

Acknowledgments

First and foremost, we would like to acknowledge the contribution of the marines, retired marines, and Marine Corps civilians whom we interviewed for this project. Respondents were assured of anonymity, but you know who you are, and this study would not have been possible without your generosity with your time and candor regarding your experiences and insights.

We can thank by name several individuals at Marine Corps Intelligence Department: Colonel Tim Oliver, who initiated the project with us; Cheryl Young, our sponsor point of contact, project monitor, and project facilitator; and the Director of Intelligence himself, Brigadier General Vincent R. Stewart. Their personal attention enabled us to gain access to and speak with a large number of Marine Corps intelligence personnel. Within RAND, we benefited from comments on interim briefs and drafts by colleagues Jim Bruce, Ben Connable, and Mike Hix. We also thank Susan Everingham and Mark Sparkman for their formal reviews and helpful suggestions for the draft monograph as part of RAND's quality assurance process. Without the efforts of RAND communication analyst Jerry Sollinger and administrative assistants Alexander Chinh and Maria Falvo, this monograph would be less well organized, less clear, and supported by fewer citations. Finally, we thank our RAND publications editorial and production team, Matthew Byrd, Carol Earnest, and Lauren Skrabala, for improving the readability and internal consistency of this monograph and seeing it through production into the final form you see here.

Errors and omissions remain the responsibility of the authors alone.

Abbreviations

ACE	Air Combat Element
BCT	brigade combat team
BfSB	battlefield surveillance brigade
C2	command and control
CE	command element
CI	counterintelligence
CIA	Central Intelligence Agency
CLIC	company-level intelligence cell
COIN	counterinsurgency
DCI	Director of Central Intelligence
DIA	Defense Intelligence Agency
DIRINT	U.S. Marine Corps Director of Intelligence
DoD	U.S. Department of Defense
DOTMLPF	doctrine, organization, training, materiel, leadership and education, personnel, and facilities
EMW	expeditionary maneuver warfare
FY	fiscal year

G-2	derived from -2 for intelligence in the Napoleonic staffing system; the intelligence staff (see also S2)
GCE	Ground Combat Element
GEOINT	geospatial intelligence
HUMINT	human intelligence
IC	U.S. Intelligence Community
I-Dept	U.S. Marine Corps Intelligence Department
IID	U.S. Marine Corps Intelligence Integration Division
INSCOM	U.S. Army Intelligence and Security Command
ISR	intelligence, surveillance, and reconnaissance
IT	information technology
JMIP	Joint Military Intelligence Program
JSF	Joint Strike Fighter
LCE	Logistics Combat Element
MAGTF	Marine Air-Ground Task Force
MARFOR	U.S. Marine Corps Forces
MARFORCOM	U.S. Marine Corps Forces Command
MARSOC	U.S. Marine Corps Forces Special Operations Command
MASINT	measurement and signature intelligence
MCCDC	U.S. Marine Corps Combat Development Command
MCIA	U.S. Marine Corps Intelligence Activity

MCIS	U.S. Marine Corps Intelligence School
MCISR-E	U.S. Marine Corps Intelligence, Surveillance, and Reconnaissance Enterprise
MCSB	Marine Cryptologic Support Battalion
MCSC	U.S. Marine Corps Systems Command
MEB	Marine Expeditionary Brigade
MEF	Marine Expeditionary Force
MEU	Marine Expeditionary Unit
MI	military intelligence
MIC	MEF Intelligence Center
MIP	Military Intelligence Program
MOS	military occupational specialty
NDS	National Defense Strategy
NGA	National Geospatial-Intelligence Agency
NIP	National Intelligence Program
NMS	National Military Strategy
NSA	National Security Agency
NSS	National Security Strategy
OEF	Operation Enduring Freedom
OIF	Operation Iraqi Freedom
PIR	priority intelligence requirement
PM IDF&D	program manager, intelligence data fusion and dissemination
PM Intel	program manager, intelligence systems

RCT	regimental combat team
RSTA	reconnaissance, surveillance, and target acquisition
S2	derived from -2 for intelligence in the Napoleonic staffing system; the intelligence staff (see also G-2)
SIGINT	signals intelligence
SRIG	surveillance, reconnaissance, and intelligence group
T/O	Table of Organization
TECOM	U.S. Marine Corps Training and Education Command
TIARA	Tactical Intelligence and Related Activities
UAS	unmanned aircraft system
UAV	unmanned aerial vehicle
USMC	U.S. Marine Corps

CHAPTER ONE

Introduction

Background

U.S. Marine Corps (USMC) intelligence is assigned mission responsibility for all USMC intelligence matters, with functions ranging from conducting intelligence collection to conducting analysis in support of operating forces in combat and deployed around the world. It also represents the Marine Corps in the U.S. Intelligence Community (IC) and supports the U.S. Department of Defense (DoD) resource allocation processes. Particularly since 2001, the USMC intelligence enterprise has demonstrated agility in tailoring its organization to meet evolving expeditionary force demands. This has resulted in a number of ad hoc arrangements, practices, and organizational structures. USMC operations include distributed operations, irregular warfare, amphibious warfare, and joint and coalition warfare. These demands, combined with the increasingly rapid pace of technological change, have challenged the organizational capability of USMC intelligence to both meet the requirements of Fleet Marine Forces in the current operating environment and ensure effective participation in the broader IC, including compliance with various IC and DoD mandates.

There are multiple reasons to review the organizational structure and design of USMC intelligence. First, it has been more than 15 years since the 1994 Intelligence Plan (the so-called Van Riper Plan) was launched in response to perceived shortcomings exposed by the Gulf War. It is an open question how many of that era's issues were effectively addressed through the implementation of the 1994 plan; further challenges have emerged since then, and others may have been created through the plan's implementation. Second, in addition to the changes wrought by the 1994 Intelligence Plan, a decade of sustained employment in Operation Iraqi Freedom (OIF) and Operation Enduring Freedom (OEF) has led to changes in the workforce and structure of USMC intelligence. Since 2006, the USMC itself has grown from 175,000 to 202,000 marines, and the number of marines with intelligence military occupational specialties (MOSs) has more than doubled since 1994.[1] Continuous counterinsurgency (COIN) operations have changed tactical support structures, and technological innovations have provided new tools and capabilities. Third, the attacks of September 11, 2001, led to reform in the larger IC, with some impact on USMC intelligence, including changed relationships within the IC and the establishment of the USMC as the IC lead for cultural intelligence. Fourth, the information environment itself has changed substantially since 1994, with different sources of information becoming available and more prevalent, new information-gathering technologies being developed, and evolving needs for and means of disseminating information and intelligence among the operating forces. Finally, with OIF concluded and the end of OEF in the foreseeable future, a new era of austerity looms. Secretary of Defense Robert Gates has already launched initiatives to reduce defense spending over the next five years.[2] Rumors have also suggested that the USMC will draw down from its current end strength of 202,000; the 2011 report of the USMC Force Structure Review Group plans for a force of approximately 186,800 active-duty marines following the conclusion of operations in Afghanistan.[3] What does this mean for USMC intelligence going forward?

The USMC Director of Intelligence (DIRINT) asked the RAND National Defense Research Institute to examine ways of aligning the organizational structures of the USMC intelligence enterprise to

---

[1]   All Marines Memo 008/07, "Marine Corps End Strength Increase," February 7, 2007.

[2]   Robert M. Gates, Secretary of Defense, "SECDEF Statement," Washington, D.C., August 9, 2010.

[3]   Headquarters, U.S. Marine Corps, *Reshaping America's Expeditionary Force in Readiness: Report of the 2010 Marine Corps Force Structure Review Group*, Washington, D.C., March 14, 2011.

efficiently and effectively carry out current and future missions and functions.

Recent History of Marine Corps Intelligence

Since the end of the Cold War, USMC intelligence has undergone significant organizational change.[4] In the early 1990s, a drastically different strategic context and fiscal environment precipitated a broad rethinking of roles and missions in the armed forces, and the USMC was no exception. A sweeping review in 1994 led to a package of significant reforms known as the Intelligence Plan, or the Van Riper Plan, after the general who played a significant role in shaping it.[5] It identified deficiencies with regard to specific disciplinary competencies, training, professional development, and tactical intelligence. The plan included a reform program based on seven fundamental principles that enshrined a commitment to tactical intelligence and professionalizing the workforce. While it ushered in significant improvements in some areas, the plan did not meet expectations in others. Progress toward meeting Intelligence Plan objectives included a growth of 56 percent in intelligence manning between 1994 and 2006.[6] It also established a career track for intelligence marines and four new entry-level training tracks for officers, organized by intelligence discipline, and it also launched efforts to improve capabilities. Yet, in the years after the plan's adoption, writers in the *Marine Corps Gazette* continued to bemoan what they saw as continued weak links between intelligence and opera-

---

[4] The recent history of the USMC intelligence enterprise is explored in greater detail in Appendix D.

[5] Paul K. Van Riper, "Observations During Desert Storm," *Marine Corps Gazette*, Vol. 75, No. 6, June 1991; All Marines Memo 100/95, "Program to Improve Marine Corps Intelligence," March 24, 1995.

[6] U.S. Marine Corps Intelligence Department, "'202K' Build Out for Marine Corps Intelligence," Washington, D.C., undated.

tions, problems with intelligence training, and a persistent "crisis of credibility" for intelligence personnel.[7]

The past two decades have also seen institutional change, both at the national and USMC intelligence levels. National-level changes included the establishment of the Office of the Director of National Intelligence in 2005 as part of broader efforts to improve coordination and integration of intelligence activities. There have been significant institutional changes in the USMC as well. In 1999, it established three intelligence battalions, one to support each MEF.[8] The next year, the Commandant established the Intelligence Department (I-Dept), raising intelligence from its previous position as a division within command, control, communication, computers, and intelligence.[9] In 2001, USMC headquarters raised the profile of the U.S. Marine Corps Intelligence Activity (MCIA) by changing it from a field activity into a command. The change to MCIA, and an expansion of its capabilities, reflected an emphasis on providing better tactical support to operators—as had been envisioned by the Intelligence Plan.[10]

For almost a decade, USMC intelligence has been an organization at war. This has posed significant challenges, but it has also offered unique opportunities. USMC responsibilities have included conventional "forced-entry" operations, counterterrorism, and COIN operations. To meet these challenges, the Secretary of Defense approved an expansion of USMC end strength to 202,000.[11] The USMC has also pursued innovative approaches to the organization of intelligence resources, such as the widely discussed distribution of intelligence below the battalion level. Recent operations have highlighted the need to bolster key areas of expertise, especially in the selec-

---

[7] E. Ennis Michael, "The Future of Intelligence," *Marine Corps Gazette*, Vol. 83, No. 10, October 1999, p. 46.

[8] R. Liebl Vernie, "The Intelligence Plan: An Update," *Marine Corps Gazette*, Vol. 85, No. 1, January 2001, p. 54.

[9] Michael, 1999, p. 46.

[10] Vernie, 2001, p. 54.

[11] F. G. Hoffman, "The Corps' Expansion," *Marine Corps Gazette*, Vol. 91, No. 6, June 2007, p. 42.

tion and training of intelligence analysts and midcareer personnel. Moreover, deciding how to capture lessons learned to retain hard-won capabilities to meet challenges beyond current operations will be a central concern for USMC intelligence as it organizes for the future. See Appendix D for a recent history.

Organization of This Monograph

Chapter Two outlines the approach that the research team used for its assessment. Chapter Three documents the current organization and manpower of the USMC intelligence enterprise. Chapter Four reviews the relevant literature on organizational design. Chapter Five uses USMC documentation as the basis for a statement of strategic intent in the form of objectives for USMC intelligence. Chapter Six outlines the issues that surfaced in the semistructured interviews that the research team conducted with a range of USMC personnel and civilians. Chapter Seven discusses organizational structure issues and makes recommendations, while Chapter Eight discusses the resolution of the issues identified in Chapter Six. Chapter Nine provides conclusions and overall recommendations. The six appendixes summarize the organizational literature reviewed for this study, the organization of Army intelligence capabilities as a point of comparison, the interview topics and questions, a recent history of USMC intelligence, current strategic guidance, and additional details about the assessment of organizational alternatives, the results of which were presented in Chapter Seven.

CHAPTER TWO

Approach

This chapter describes our approach to this study. This undertaking involved five mutually supporting and related strands of research effort:

- a review of the literature on organizational design
- a review of documents for and about USMC intelligence
- semistructured interviews of personnel in USMC intelligence organizations
- analyses of these data
- the development and assessment of organizational alternatives for Marine Corps intelligence.

In this chapter, we discuss each of these areas in turn.

Literature on Organizational Design

To develop a framework to assess the organizational baseline of USMC intelligence and to evaluate alternative courses of action and identify issues of concern, we scoured the existing literature on organizations. The sources we reviewed are listed in Appendix A. Using both formal academic organizational theory sources and concepts in the broader business literature, we identified a host of models, schemes, frameworks, and approaches with which to study organizations. The review also included RAND reports on military organizations that outlined organizational assessment methods and conclusions that are specific to the military.

From each source, we collected information about organizational characteristics, features, or criteria that the various contributors identified as relevant to organizational design or analysis. We then determined why they highlight those aspects (i.e., their theory), with a focus on characteristics of organizations, such as strategy, missions, environment, leadership, and technology. We also examined basic organization types, including their functional structures, divisional structures, and organizational matrices. Findings from the synthesis of this literature review are reported in Chapter Four.

After selectively reviewing the literature, we chose to rely primarily on the organizational work of Burton, DeSanctis, and Obel.[1] Specifically, they provide a methodological approach to organizational assessment that relies on the concept of organizational "fit." A structural form should fit with the goals of the organization and its environment. Moreover, other aspects of an organization should then fit with the chosen structural form. Given our research charge, we did not pursue the full Burton analysis, but we used his approach to inform our assessment of structure as it fits with the goals and environment of USMC intelligence. While we relied primarily on Burton, the remaining literature provided a rich source of detail for implementing certain organizational structures, such as a matrix. Finally, as discussed in Chapter Four, the use of the design literature was tempered by the research team's experience with prior organizational studies.

Marine Corps and Marine Corps Intelligence Strategy, Plans, and Doctrine

With a foundation for the assessment of organizational design in place, we sought to contextualize USMC intelligence within our organizational design framework. We started this process with a review of material on the USMC and USMC intelligence. The goal of this review was twofold. First, we sought to understand the current (and

---

[1]   Richard M. Burton, Gerardine DeSanctis, and Børge Obel, *Organizational Design: A Step-By-Step Approach*, Cambridge, UK: Cambridge University Press, 2006.

recent historical) organization of and challenges facing USMC intelligence. Second, we sought to understand current and evolving strategic goals and guidance for the USMC in general and USMC intelligence specifically. This review contributed to the overviews presented in Chapter Three, which examines current organization, and Chapter Five, which addresses strategic intent. The review of strategic intent, in turn, contributed to our broader analysis of organizational alternatives (Chapter Seven and Appendix F) and helped establish a scheme for prioritizing the issues identified in Chapter Six.

Specifically, our review of the strategic guidance included the following:

- The Marine Corps Intelligence Plan (1994)
- *National Military Strategy* (2004)
- The Marine Corps Intelligence, Surveillance, and Reconnaissance Roadmap (2006)
- *A Cooperative Strategy for a 21st Century Seapower* (2007)
- *National Defense Strategy* (2008)
- *Marine Corps Vision and Strategy 2025 (2008)*
- *Vision 2015* by the Office of the Director of National Intelligence (2008)
- *National Intelligence Strategy* (2009)
- *Quadrennial Intelligence Community Review* (2009)
- *National Security Strategy* (2010)
- *Quadrennial Defense Review Report* (2010)
- *The Marine Corps Intelligence, Surveillance, and Reconnaissance Enterprise (MCISR-E) Roadmap* (2010)
- *Marine Corps Operating Concepts* (2010).

In addition to these strategic-level documents, we reviewed USMC doctrine on intelligence, MOS roadmaps for intelligence, and a variety of lessons learned and observations—both formal (i.e., from the USMC Center for Lessons Learned) and informal (i.e., as recorded in the *Marine Corps Gazette* and other periodicals). It is interesting to note that many of the recent innovations in Marine Corps intelligence were, in fact, pioneered in the infantry battalions before spreading into the intelligence function.

Interviews

A review of USMC intelligence doctrine, documents, and strategy could carry our understanding of the organization of USMC intelligence only so far. To collect additional data on the nature and functioning of USMC intelligence organizations for use in our organizational design framework, we conducted a series of interviews with USMC personnel and civilians. These interviews proved foundational for our understanding of the current USMC intelligence organization, confirming our document-based evaluation of strategic intent. Feedback from the interviews was also essential to our holistic analysis, which involved making assessments of organizational fit (see Chapter Seven). Furthermore, as discussed later, the interviews provided a trove of "gripes," many of which, when distilled and synthesized, contributed to a list of issues of potential concerns. While many of these issues were not directly structural or organizational, their enumeration and prioritization should still be valuable to the USMC intelligence leadership as it seeks to improve the enterprise. Many of these issues can potentially be resolved by the organizational alternatives recommended in Chapter Seven; some suggestions for progress on nonstructural issues are offered in Chapter Eight.

We sought to interview respondents who were broadly representative of the breadth of USMC intelligence, including the supporting establishment, the command element, and the combat elements.[2] We identified potential respondents based on structural position (who is currently serving in which positions in which commands) and by referral (i.e., we usually asked respondents to recommend others whom we could speak with regarding the issues under discussion). In all cases, interviews were voluntary and contingent on our ability to contact the

---

[2] Ideally, we would also have interviewed an extensive array of intelligence customers in the USMC, but that did not prove feasible in the time available.

individuals, as well as their availability during the days of our site visits
or for follow-up phone interviews. All interviews were conducted on a
not-for-attribution basis in order to promote candor.

Interview Participants

We interviewed a total of 120 respondents in 65 interview sessions. The
sessions included between one and eight respondents; two was typical. Respondents were a mix of military and civilian personnel. Of the
interviewees, 60 were officers, 30 enlisted, and 30 civilians. In terms
of experience, 80 were grades O-4, E-7, and comparable civilian and
above, while 40 could be characterized as junior.

Interview Topics and Questions

Our list of interview topics and questions included 82 questions divided
into 16 categories derived from our review of the literature on organizational design (described in Chapter Four). The questions fell into the
following categories:

- the interviewee and his or her affiliation
- basic organizational structure
- organization-level goals
- relationships with other organizations
- complexity and unpredictability of the environment
- products
- innovation
- internal coordination
- distribution of operations and decisionmaking
- organizational knowledge, information flow, and information
  technology (IT)
- task design
- formalization and centralization
- process and competitive advantage
- personnel
- delegation and uncertainty avoidance
- organizational climate and culture.

In practice, 82 questions are too many to pose in an interview of reasonable length. Therefore, we provided the interview topics and questions to many of the subjects as read-ahead material and then used the list as a reference to help guide and focus our discussion. With other respondents, we targeted specific questions based on their positions and relevant experience. Interviews ranged from 30 to 120 minutes, depending on time available, the intensity of the discussion, and the number of appointments scheduled during a particular site visit. The list of interview topics and questions is presented in its entirety in Appendix C.

Data Analysis

The interviews, once the notes were transcribed, produced a very rich database. In addition to the valuable holistic insights drawn from conducting the interviews and reviewing the transcripts (which supported our understanding of the "as-is" organization of USMC intelligence and our analysis of organizational alternatives), we sought to conduct a more structured analysis using software for the management and coding of qualitative data.

We used ATLAS.ti (version 6) software to classify responses to our interview questions into categories.[3] ATLAS.ti enables the analyst to select any segment of a document (in this case, notes from the interviews) and assign it to one or more user-defined categories. The segments that the user selects are referred to as "quotes," and the user-defined categories are referred to as "codes." Quotes were coded at a broad level to provide a general picture of the topics discussed. A single quote could be coded into multiple categories. For example, the following statements were selected as one "quote" and coded into several categories, including "innovation," "mission/strategy/planning," and "manpower and staffing."

---

[3] ATLAS.ti is a product of ATLAS.ti Scientific Software Development GmbH.

Much of the work is reactive. But we get a lot of mileage (a lot of work done), especially relative to our size. We don't have much time to look ahead, so we set aside time to do that. Try to free up one or two people to do the big things, like the roadmap. The [executive steering advisory group] is one of the ways we look ahead.

This group of people tries to do the day-to-day work and leave the big thinkers the ability to stay free and think about the future.

[The] MC intel community is small, everybody knows each other.

We coded the transcripts from the 65 interview sessions, comprising 120 individual interview respondents. We derived 24 codes from our review of the organizational literature that informed our interview questions and from themes that emerged in interviewees' responses. The substantive codes included the following:

- agility
- authority/grade/rank
- bureaucracy
- career progression
- combat operations (relationship with)
- competitive advantage
- culture
- enterprise
- financial resources/budget
- information and communication technology and tools
- innovation
- intra- and interorganizational relationships
- knowledge management
- location of intelligence organizations (geographic location)
- manpower and staffing
- mission/strategy/planning
- organizational structure
- peacetime and in-garrison activities
- personality

- reachback and reachforward
- standards
- training
- value of intelligence (how intelligence is perceived by others)
- other.

In addition to coding substantive themes, we classified comments in terms of the organization being referenced by the interviewee. These codes included the following:[4]

- ACE (Air Combat Element)
- CE (command element)
- CI/HUMINT (counterintelligence/human intelligence)
- G-2 (intelligence staff)
- HQ (headquarters)
- I-Dept
- infantry battalion
- intelligence battalion
- logistics battalion
- MAGTF (Marine Air-Ground Task Force)
- Marine Corps Forces (MARFOR, e.g., MARFOR Command [MARFORCOM], MARFOR Pacific, MARFOR Systems Command)
- MCCDC (USMC Combat Development Command)
- MCIA
- MCIS (USMC Intelligence School)
- MEF (Marine Expeditionary Force)
- MEU (Marine Expeditionary Unit)
- radio battalion
- three-letter agencies (e.g., Central Intelligence Agency [CIA], National Security Agency [NSA], National Geospatial-Intelligence Agency [NGA]).

---

[4] We originally employed a broader set of organizational categories but dropped many because of low frequency of reference.

Once the interview notes were coded, we reviewed the quotes for each code to identify themes pertaining to the key topics of interest and specific issues of concern to multiple respondents. The resulting list of 48 unresolved issues is discussed in Chapters Six and Eight.

While that list includes many issues that do not pertain directly to organizational structure, we include all of the identified issues here because of their possible utility to those seeking to reform and improve USMC intelligence. Rather than simply providing the USMC with a simple list of issues to consider or address, we sought to prioritize them, suggesting which issues should be of greatest concern. However, we were unable to do so on the basis of frequency of mention or occurrence in our interview sample for several reasons. First, our sample is not a probability sample and is not proportionally representative of a discernable and distinct portion of the broader USMC intelligence enterprise. Our sample includes representatives from all of the organizational elements that we sought, but not in fixed proportions. Second, there were variations in the number of personnel available at each of the organizations during our site visits, and we added additional respondents through direct referral from interviewees. Third, although our respondent sample was broadly representative of USMC intelligence organizations, the actual interviews varied in both duration and in the specific questions asked or topics discussed. In general, interview respondents are inclined to mention and discuss areas highlighted in the line of questioning and less inclined to mention areas not asked about. For these two reasons, we refrained from attempting to make inferences based on the frequency with which an issue is mentioned, beyond requiring that an issue be mentioned by more than one respondent before we considered it a confirmed issue and not just a single respondent's pet peeve.

Since we could not prioritize the issues based on their appearances in the interview responses, we sought an external referent. Fortunately, our broader organizational assessment effort required that we identify clear goals and objectives for USMC intelligence organizations. We derived a set of seven organizational objectives from our review of the existing strategic guidance. This effort, supported by our review of USMC doctrine, is described in greater detail in Chapter Five. We assessed the potential threat posed by each of the 48 issues to each of the seven objectives, if left unaddressed. We then weighted the level of threat and the objectives themselves, resulting in priority scores and ranks for the 48 interview-derived issues. This analysis is presented in Chapter Six.

Development and Assessment of Alternative Structures

Our review of USMC intelligence-related documentation and the interviews we conducted allowed us to describe and enumerate the existing organizational structure on several levels (see Chapter Three). Our review of the literature on organizational design provided appropriate terms to both describe and assess those baseline organizations. Using a framework developed from the literature review (and described in Chapter Four), we assessed the "organizational fit" of the design characteristics for four organizational levels of USMC intelligence. Specifically, we assessed the organizational fit of the I-Dept, MCIA, MEF intelligence structures (specifically, the intelligence battalion and radio battalion, and intelligence structures throughout the combat elements, including the Ground Combat Element (GCE), the Air Combat Element (ACE), and the Logistics Combat Element (LCE). These latter levels of organization are division, wing, or logistics group units and below. Where the current organizational design did not fit with organizational objectives and missions, we identified alternative organizational structures based on comments from the interviews or from the literature on organizational design. We then assessed these alternatives using the same criteria as for the baseline organizational structures, as described in Chapter Seven; methodological details are presented in Appendix F.

CHAPTER THREE

Current Organization of Marine Corps Intelligence

This chapter describes the current organization of USMC intelligence. In Appendix B, we have included a discussion of Army intelligence for those who are inclined to ask the obvious comparative question, "How does the Army organize for intelligence?"

USMC intelligence manpower and units are housed at the headquarters level (called the supporting establishment) and in operational forces consisting of several forces-level commands and their subordinate units. Operational emphasis is placed on the MAGTF, which comprises forces organized by task under a single commander and is structured to accomplish a specific mission. The MAGTF has four core elements: the CE, GCE, ACE, and LCE.[1]

The Organization of Marine Air-Ground Task Forces

The four core elements are present in the three types of MAGTFs, the largest of which is the MEF, typically a three-star command. There are three MEFs, each with an intelligence battalion and a radio battalion. The intelligence battalion is composed of a battalion headquarters, a headquarters company, a production and analysis company, a production and analysis support company, a CI/HUMINT company, and

---

[1] This summary is drawn from a number of sources, including USMC websites; Headquarters, U.S. Marine Corps, *Organization of Marine Corps Forces*, Washington, D.C., Marine Corps Reference Publication 5-12D, October 13, 1998; and the structure provided by the Total Force Manpower Management System.

a CI/HUMINT support company. A number of sections and teams make up the company-level units. In total, there are about 75 officers and 600 enlisted personnel in a battalion. Of this number, about 550 have an intelligence MOS, either in the 02 or 26 career field, with the vast majority in 02.[2]

The intelligence battalion is responsible for planning, producing, and disseminating intelligence, as well as providing CI support to the MEF CE. The radio battalion handles both signals intelligence (SIGINT) and electronic intelligence.

Second in size to the MEF is the Marine Expeditionary Brigade (MEB), which is capable of conducting missions across a full range of military operations. The smallest type of MAGTF is the MEU, which is structured as an expeditionary quick-reaction force, ready to respond immediately to any crisis. The MEF, MEB, and MEU are part of the CE.

The elements are the operating forces at the division, wing, and logistics levels and below. Intelligence personnel are part of the manpower component in each of these units, ranging from about 60 at the division level to around ten in an infantry battalion and even fewer in a logistics or engineer battalion. The ACE is the core element of a MAGTF that is task-organized to conduct aviation operations. The ACE is usually composed of an aviation unit headquarters and various other aviation units or their detachments. The GCE is the core element of a MAGTF that is task-organized to conduct ground operations. It is usually constructed around an infantry organization but can vary in size from a small ground unit of any type to one or more divisions that can be independently maneuvered under the direction of the MAGTF commander. The LCE is the core element that is task-organized to provide the combat service support necessary to accomplish the MAGTF mission. The combat service support element varies in size from a small detachment to one or more force service support groups.

---

[2] MOS codes have four digits, with the first two denoting the occupation (or career) field. An MOS beginning in 02 is in the intelligence career field; an MOS beginning in 26 is in the SIGINT/ground electronic warfare career field.

The Organization of the Intelligence-Supporting Establishment

The USMC intelligence-supporting establishment consists of head-quarters-related organizations, and a key layer is the I-Dept. The I-Dept is responsible for overseeing policy, plans, programming, budgets, and personnel supervision in USMC intelligence. The department, headed by the DIRINT, supports the Commandant of the Marine Corps. The DIRINT is assisted and supported by the Assistant Director of Intelligence and staff. The Intelligence Plans Division is responsible for oversight of intelligence requirements and capabilities planning, development, and integration. In addition, the division coordinates geospatial intelligence (GEOINT), SIGINT, meteorology and oceanography, and electronic warfare programs. The Intelligence Operations Division, on the other hand, provides intelligence support to headquarters and CI, SIGINT, and HUMINT management. The division is also responsible for intelligence estimates, which fall under the Intelligence Estimates Branch. Among its various tasks, the Intelligence Estimates Branch compiles and disseminates completed intelligence reports to the Commandant and principal staff officers. It also acts as a liaison with other national and departmental intelligence services, such as the CIA and the Defense Intelligence Agency (DIA), and participates in the formulation of U.S. Joint Chiefs of Staff documents pertaining to intelligence matters.

MCIA, another part of the supporting establishment, provides a key reachback resource for marines and a locus for analytical efforts. It consists of staff elements and three military components: a production and analysis company, the Marine Cryptologic Support Battalion (MCSB), and a CI/HUMINT support company. The production and analysis company provides imagery support as well as both all-source and cultural intelligence. The MCSB provides regionally focused support to SIGINT analysis and coordinates with the radio battalions. The CI/HUMINT support company is responsible for service-level HUMINT collection management.

The Intelligence Integration Division (IID) of the MCCDC, the PM Intel (program manager, intelligence systems), the PM IDF&D

(program manager, intelligence data fusion and dissemination), and the intelligence schools account for the remaining key elements of the intelligence-supporting establishment.

Manpower Resources

Growth

Authorizations for USMC intelligence personnel increased in the early 1990s and again after September 11, 2001, as the USMC itself grew to 202,000 personnel. The intelligence structure has grown from 478 officers and 2,642 enlisted in 1994 to more than 1,050 officers and 5,170 enlisted. In fiscal year (FY) 2000, intelligence personnel represented 2.6 percent of the USMC. This increased to 3.6 percent in FY 2009, a figure comparable to the density of intelligence personnel in the Army. See Appendix B for a description of how the U.S. Army organizes for intelligence.

Distribution by Organization Level

The CE has the majority of all intelligence manpower—officers and enlisted personnel in occupational fields 02XX and 26XX. (Civilian personnel and contractors are not included in these numbers.) Fifty-five percent of military intelligence (MI) personnel are in the CE, largely in the intelligence and radio battalions. Another 20 percent is in the supporting establishment, mainly at MCIA. The combat elements, the GCE, ACE, and LCE, account for 13, 11, and 1 percent of manpower, respectively. These data are shown in Figure 3.1.

Distribution by Grade

In terms of intelligence authorizations, the more experienced officers are in the supporting establishment and CE, while the less experienced are in the three combat elements. More than 70 percent of the authorizations for lieutenants are in the combat elements, while over 65 percent of captains are authorized in the CE and supporting establishment. Among majors and lieutenant colonels, more than 90 percent are in the CE and supporting establishment (see Figure 3.2).

Figure 3.1
Distribution of Manpower, by Level

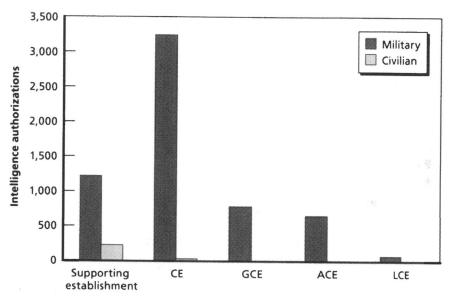

SOURCE: Data from the USMC Total Force Structure Management System, as of
April 12, 2010.
RAND MG1108-3.1

There is another factor that plays into this authorization grade distribution: the actual grade makeup of the inventory of intelligence officers. Because there has been recent growth in the number of officers, and due to the closed nature of the military personnel system, the inventory of intelligence officers is more junior than that of other occupations. USMC intelligence officers are overrepresented in the O-1, O-2, and O-3 grades and underrepresented at the O-4 and O-5 level, compared to infantry and artillery officers (see Figure 3.3).

Moreover, intelligence officers have, on average, less experience at the grade of O-3 than their counterparts. Time will eventually resolve both these issues, given comparable promotion and retention outcomes.[3]

---

[3] This is an assumption based on experience with other closed-entry systems in which it takes time for new additions of large numbers of junior personnel to move through the system and gain experience in grade. Our expectation is that when the recent rapid growth

Figure 3.2
Distribution of Manpower, by Grade

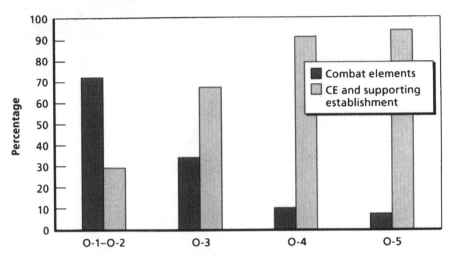

SOURCE: Data from the USMC Total Force Structure Management System, as of
April 12, 2010.
RAND *MG1108-3.2*

Figure 3.3
Officer Grade Distribution

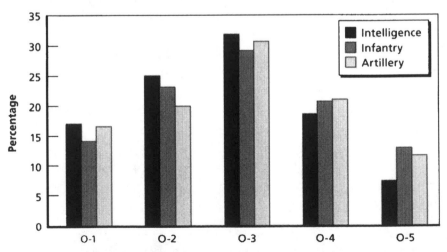

SOURCE: Data from the USMC Total Force Structure Management System, as of
April 12, 2010.
RAND *MG1108-3.3*

The enlisted grade distribution by organization level is similar to that of officers, with more than 80 percent of E-5 and E-6 authorizations and nearly 90 percent of E-7, E-8, and E9 authorizations in the CE and supporting establishment.

Use of Inventory in a Deployed Setting

Because the USMC typically task-organizes, we also examined the organizational location for deployed intelligence marines. At issue is this question: Are large numbers of intelligence personnel located in regimental and below areas of operation, or are they in more centralized locations, such as Camp Leatherneck? A snapshot on one day in July 2010 showed that one-third of all intelligence marines deployed to Afghanistan were outside a central location, while two-thirds were centrally located. This varied significantly by organizational level. For example, 75 percent of GCE personnel were outside a central location (division staff being the major exception), but all ACE personnel were in a central location. Seventy-five percent of the LCE was centrally located, as was 80 percent of the CE and 70 percent of the intelligence and radio battalions. In essence, while some of the CE capability is in the regimental and below areas, it is not a large percentage.

Marine Corps Intelligence Units of Analysis

Figure 3.4 summarizes the intelligence structure discussed earlier. Our study focused on four organizational levels: (1) the I-Dept (DIRINT and immediate staff), (2) MCIA, (3) the intelligence and radio battalions, and (4) the combat elements, primarily the GCE.

---

in the number of intelligence personnel (far more rapid than in most USMC occupational fields) is assimilated and matures, the distribution of personnel by grade will come to more closely resemble that of other occupational fields. In other words, over time, officer grade distribution will normalize, with the distribution of intelligence officers looking much more similar to the distributions in infantry and artillery. This assumption and the rate at which it is being realized could be tested by tracking the progress of the data informing Figure 3.3 over time. This expected trend toward similar experience levels could fail to materialize if retention of intelligence marines significantly lags retention in other USMC occupational fields.

Figure 3.4
Structure of Marine Corps Intelligence and the Four Organizational Levels Analyzed

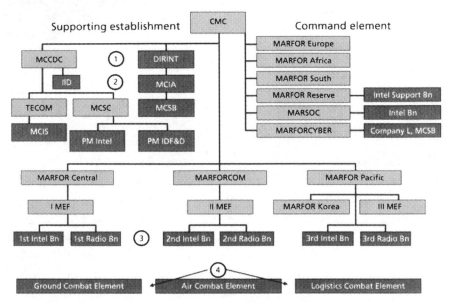

NOTE: Bn = battalion. MARSOC = USMC Forces Special Operations Command.
MCCD = USMC Combat Development Command. MCIA = USMC Intelligence Activity.
MCSC = USMC Systems Command. MEF = Marine Expeditionary Force.
TECOM = USMC Training and Education Command.
RAND MG1108-3.4

CHAPTER FOUR

Literature on Organizational Design and Analytic Framework

The RAND team brought to the project considerable prior experience with organizational design and theory. In our research proposal, we included a model of the relationship between organizational structural components and an organization's inputs and outputs. This model, outlined in Figure 4.1, is based on work by the National Research Council.[1] It seemed appropriate because, as many authors have pointed out, the success of an organizational structure can depend on a number of factors. Important among them is an organization's culture. A new structure may perform much as the old one did because the culture impedes changed effectiveness or efficiency. Cultural inertia can limit change, and the USMC is an organization with a strong, old, uniform culture. However, it is also likely that its strong culture, especially with respect to supporting the combat elements, can be leveraged to promote certain changes.

As noted in Chapter Two, we conducted a substantial review of additional literature on organizational design and theory, looking for new developments or additional insights beyond our preexisting expertise. A list of the materials reviewed is included in Appendix A. This undertaking contributed to our analysis in three ways. First, it confirmed our baseline input-output model (shown in Figure 4.1) as both reasonable and consonant with much of contemporary thinking on organizational design. Second, it allowed us to systematically identify the wide range of categories of organizational data discussed in the

---

[1] See Druckman, Daniel, Jerome E. Singer, and Harold Van Cott, eds., *Enhancing Organizational Performance*, Washington D.C.: National Academies Press, 1997.

Figure 4.1
Effects of Organizational Structure on Inputs and Outputs

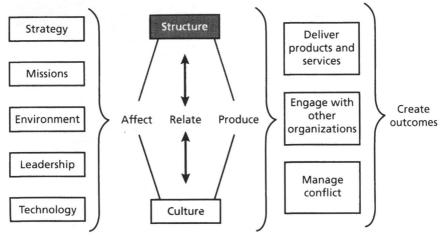

NOTE: The left side of the model is described as "what the organization needs," the center as "what the organization is," and the right as "what the organization does."
RAND *MG1108-4.1*

literature and to develop our interview topics and questions to include all likely relevant categories (see Appendix C). Third, we used the analytical approach of "organizational fit" as suggested in work by Burton et al. and incorporated it into our analytic framework and approach.[2]

Organizational Fit

Our framework is based on the synthesis and consolidation of a considerable list of organizational characteristics of interest from the design literature. The core framework, many of the initial categories, and the notion of organizational fit come from Burton et al.[3] However, much of the organizational design literature reviewed in Appendix A contains similar methods of analysis and structural prescriptions, and

---

[2]   See Burton, DeSanctis, and Obel, 2006.

[3]   Burton, DeSanctis, and Obel, 2006.

these sources provided details on the benefits and costs of different structural forms.

We present the key aspects of our organizational framework here, along with the nested logical nature of the various elements.

1. In terms of design, one does not design an entire organization globally because key design elements (e.g., goals, strategy, environment) might be different for each piece of the organization.
2. Each of the design elements fits better or worse with other design elements. At each step of the application of this method, one can assess "fit" (e.g., strategy fits with goals and environment) or "misfit" (it does not).
3. One can analyze each piece of the organization "as is" (in its current state) and ask where would it should be (e.g., in terms of goals or strategy) if its elements do not fit in their current alignment.

The Burton methodology works through sequential steps by answering a series of diagnostic questions in particular areas. The first step is goals, the next is strategy, the third is environment, and next is structure, followed by process, people, coordination, and control.

Hierarchical Criteria

Our framework calls for the assessment of (and an assessment of the fit of) four elements of each part of an organization: goals, strategy, resources and authority, and environment. Each varies in two dimensions. In this section, we review the questions that facilitate such an assessment.

Organizational goals focus on effectiveness or efficiency (or a combination of both). The core diagnostic question is, "Are your goals focused on the product (effect) or the process (efficient)? This can be further sharpened by asking, "Is your focus the customer or the institution?"

Strategy has many components, but for assessing organizational fit we need only consider the balance between exploration and exploitation in the organization's strategy. The core diagnostic question asks, "Is your strategy to take initiative (explore) or play by the rules (exploit)?" Two related questions can help refine the assessment of strategy: "Does value depend on expertise or collaboration?" and "Do you need centralized or decentralized decisionmaking?"

An organization's ability to pursue its goals and follow its strategy depends on resources and authorities. Several diagnostic questions offer support here: "Do you have sufficient resources, including manpower and authority, to employ them?" "Do you have experienced or inexperienced personnel?" and "Do you have the necessary infrastructure to support strategy and goals?"

Finally, there is the environment in which the organization operates. While goals, strategy, and resources might all be changed by changing the organizational design, environment is dictated by context. If other key organizational elements do not fit with the environment in which an organization operates, change is likely called for.

Environment varies in two dimensions: complexity and predictability. Complexity is the number of factors in the environment to be considered and their interrelatedness. Predictability is the knowledge one has about these factors. Thus, the core diagnostic question asks, "Is your work environment complex and predictable?" Nuance can be captured by asking two supporting questions: "Can you predict your schedule and demands?" and "Do you face a broad or narrow range of demands?"

Our initial assessment of the organizational emphasis for each of four levels of organization (I-DEPT, MCIA, MEF intelligence and radio battalions, and combat elements) is shown in Figure 4.2.

Basic Organizational Structures

The existing literature on organizational design describes four basic structural alternatives with a host of minor variations. The four struc-

Figure 4.2
Assessment of Organizational Emphasis

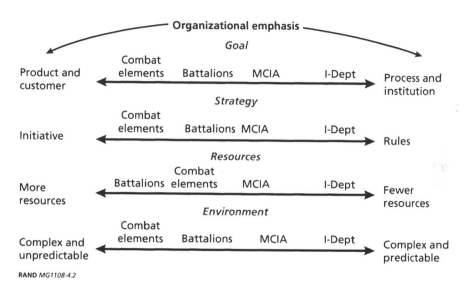

tures are simple, functional, divisional, and matrix. Figure 4.3 illustrates them graphically.

A simple structure is one in which there is an owner, supervisor, manager, or leader and a group of workers. This form is typical of a small, private business. There is no hierarchy because the owner reports to no one. This form is not likely to exist in a military environment because, even at the lowest levels of organization, such as a team or squad, there is a vertical and horizontal hierarchy to consider.

The functional structure, however, is typical of many organizations. The "functions" can represent a number of different dimensions. In this case, we represent the functions as the intelligence expertise that people bring to the organization. USMC intelligence is traditionally largely organized by function. Another example is Headquarters, USMC, where the various deputy commandants represent functional areas such as manpower, aviation, and logistics.

The divisional structure represents self-contained, independent, decisionmaking units. These units may be allocated by customer or geography. For example, in private enterprise, one unit may target

Figure 4.3
Four Basic Organizational Structural Options

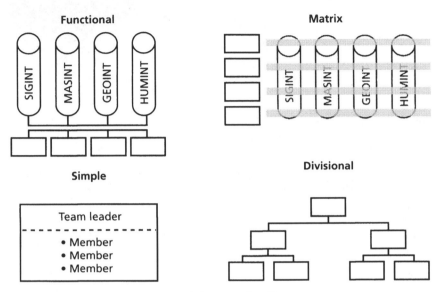

NOTE: MASINT = measurement and signature intelligence.
RAND *MG1108-4.3*

wholesale markets while another targets retail customers. Another typical structure is determined by geography, and there may be separate North American, European, and Asian divisions. In the military, this is the basis of structure for the geographic unified commands. Moreover, it is the basic structure of the ground combat elements, in which an infantry battalion or regiment has a geographic area of operations and is provided with the functional resources internally to accomplish its mission.

The matrix form of organization is a combination of the functional and divisional structures with overlapping "ownership" of resources. For example, personnel are hired, trained, promoted, and separated by one part of the matrix while the other part uses them for business- or mission-related purposes. In essence, one part of the matrix focuses on inputs while the other focuses on outputs or customers. This is a difficult form to manage because individuals report to two bosses. The matrix could, in fact, tilt toward more control by the business side or more by the functional side.

There are several ways to implement a matrix form of organization. One suggested by several organizational designers is called a front-back hybrid matrix.[4] In this form, the front end is a customer-facing unit organized by geography, customer segments, or both. The back end is organized around business units and large-scale functions. This is a dual structure in which both halves are multifunctional units. This form achieves customer responsiveness in the front and global scale in the back. The difficulty is in ensuring that the front and back are linked. This form can also be structured to provide customers with a consistent point of contact that understands their missions and needs. This habitual relationship can increase organizational effectiveness in the eyes of the customer.

The military uses the matrix structure in forms such as general support, direct support, attached, and assigned.[5] In general support,

---

[4] See, for example, Jay R. Galbraith, *Designing Matrix Organizations That Actually Work: How IBM, Procter & Gamble and Others Design for Success*, San Francisco: Jossey-Bass, 2009, or Edward E. Lawler, *From the Ground Up: Six Principles for Building the New Logic Corporation*, San Francisco: Jossey-Bass, 1996.

[5] The terms *organic, assign, attach, direct support*, and *general support* refer to specific command relationships. In Army Field Manual 101-5-1/Marine Corps Reference Publication 5-2A, a *command relationship* is defined as the "degree of control and responsibility a commander has for forces operating under his command." *Organic* is defined as "[a]ssigned to and forming an essential part of a military organization. Organic parts of a unit are those listed in its table of organization for the Army, Air Force, and Marine Corps, and are assigned to the administrative organizations of the operating forces for the Navy." *Assign* is defined in two parts as follows:

1. To place units or personnel in an organization where such placement is relatively permanent, and/or where such organization controls and administers the units or personnel for the primary function, or the greater portion of the functions, of the unit or personnel

2. To detail individuals to specific duties or functions where such functions are primary and/or relatively permanent.

Attach is the "placement of units or personnel in an organization where such placement is relatively temporary." *Direct support* refers to a "mission requiring a force to support another specific force and authorizing it to answer directly the supported force's request for assistance." *General support* is defined as the "support which is given to the supported force as a whole and not to any particular subdivision thereof" (Headquarters, U.S. Department of the Army, and Headquarters, U.S. Marine Corps, *Operational Terms and Graphics*, the functional side typically provides regional support to a number of organizations and tends to have greater control. In direct support, a functional unit is allocated to a particular supported unit with shared control of the resource between the supported and supporting units. For general and direct support, the military typically establishes habitual relationships between supported and supporting units. The same units train together in peacetime and deploy as needed. In attachment, control shifts toward the mission side and away from the functional side, which still has responsibility for providing trained and experienced personnel. In assignment, the matrix is, in fact, discarded and the functional element becomes part of the divisional structure. However, while the functional element may no longer have control of the resource, that element may still have influence over the training and experience of the functional personnel. Maintaining influence in the absence of hierarchical authority is not a simple management challenge.

Which of these structures, or variations of them, are the best forms for an organization? It depends on goals, strategy, resources and authorities, and environment.

Each of the four options may be optimal for different sets of characteristics corresponding to the different organizational elements identified here. For example, as shown in Figure 4.4, organizations whose resources include many highly skilled individuals are likely to be better off with a functional or matrix-type structure; organizations with an emphasis on the effectiveness of their product or its delivery are optimally aligned with matrix or divisional structures. Only when relatively limited expertise is required and goals emphasize efficiency is a simple organizational design appropriate.

Chapter Seven uses these basic structures to characterize the "as-is" structure of the four levels of organization we are assessing here, as well as to suggest change based on goals, strategy, resources, and environment for those levels.

First, however, we review another important consideration in structure determination: strategy or strategic intent. Following that

---

Washington, D.C., Field Manual 101-5-1/Marine Corps Reference Publication 5-2A, September 30, 1997).

Figure 4.4
Notional Mapping of Structural Alternatives with Different
Organizational Characteristics

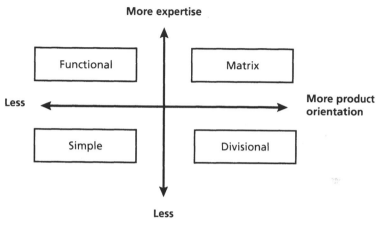

discussion, Chapter Six identifies the issues of concern with regard to the current functioning of the USMC intelligence enterprise that might affect structural decisions.

CHAPTER FIVE

Strategic Intent and Organizational Assessment: USMC Intelligence Strategy, Plans, Doctrine

When organizational leaders express a new strategic intent, they need to align the organizational structure with the new direction while accommodating history and resources. Thus, structure becomes an instrument for executing organizational strategic intent. The organizational design of USMC intelligence can be viewed through the lens of changing strategic intent and emerging strategic intent as seen in *MCISR-E Roadmap* and *Marine Corps Operating Concepts*.[1] This chapter distills existing strategic guidance into seven objectives and connects them to the organizational design characteristics described in Chapter Four.

Strategic Intent

In March 1994, the USMC announced a program to improve its intelligence collection, the Intelligence Plan (also known as the Van Riper Plan). Central to this effort was a mission statement: "Provide commanders, at every level, with tailored, timely, minimum essential intelligence, and ensure that this intelligence is integrated into the operational planning process."

The announcement identified six fundamental deficiencies to be overcome. First among these deficiencies was an inadequate doctrinal

---

[1] U.S. Marine Corps Intelligence Department, *The Marine Corps Intelligence, Surveillance, and Reconnaissance Enterprise (MCISR-E) Roadmap*, Washington D.C., April 28, 2010; Deputy Commandant for Combat Development and Integration, U.S. Marine Corps, *Marine Corps Operating Concepts*, 3rd ed., Quantico, Va., June 2010.

foundation. To solve that problem, a functional concept for intelligence was developed that laid out principles that are essential in ensuring effective intelligence support to operations. These principles are as follows:

- The focus is tactical intelligence.
- The intelligence focus must be downward.
- Intelligence drives operations.
- The intelligence effort must be directed and managed by a multi-discipline-trained and experienced intelligence officer.
- Intelligence staffs use intelligence; intelligence organizations produce intelligence.
- The intelligence product must be timely and tailored to both the unit and its mission.
- The last step in the intelligence cycle is utilization—not dissemination.

Since the program was announced, the USMC has created doctrinal publications and taken major steps to address other identified deficiencies, such as a lack of career progression for officers and insufficient tactical intelligence support.[2]

Most recently, three publications have featured an updated doctrine and strategic intent for USMC intelligence. These are *Marine Corps Vision and Strategy 2025, Marine Corps Operating Concepts, and the MCISR-E Roadmap*.[3] These documents, coupled with the organizational literature reviewed in Appendix A, form the basis for our assessment of USMC intelligence objectives. They also reveal the criticality of current obstacles and provide a means to judge the value of different organizational courses of action to mitigate those issues.

Vision and Strategy 2025 details the vision for the future and a plan for enacting it. The USMC will be a force able to act with unprec-

---

[2] See Appendix D for a recent history of Marine Corps intelligence.

[3] Commandant of the Marine Corps, *Marine Corps Vision and Strategy 2025*, Arlington, Va.: Office of Naval Research, 2008; Deputy Commandant for Combat Development and Integration, 2010; U.S. Marine Corps Intelligence Department, 2010.

edented speed and versatility in austere conditions against a wide range of adversaries. *Marine Corps Operating Concepts* presents overarching operating concepts, including the USMC's role in national security and proposed enhancements to MAGTF operations in such core areas as engagement, crisis response, and power projection. The document is intended to be read, discussed, and challenged.

The *MCISR-E Roadmap* is an appendix of the USMC Service Campaign Plan.[4] It recognizes that, to remain effective, USMC intelligence must evolve and adapt to both the changing demands of the modern battlefield and the capabilities provided by advances in technology. The service intelligence operating concept is "the synergistic integration of all Service ISR elements into a single capability or system that is networked across all echelons and functional areas including the operating forces, supporting establishment, systems and personnel in order to achieve superior decision making and enhance lethality."[5] Appendix E of this monograph describes these strategy documents in greater detail, along with strategic guidance available to the broader U.S. Intelligence Community (IC).

Objectives for the Organization of Marine Corps Intelligence

We derived seven broad objectives for the organization of USMC intelligence from the functional concept in the original 1994 Intelligence Plan, recent Marine Corps publications, and the broader organizational design literature. To this end, we extracted and decomposed the strategic intent that was explicit and implicit in the Intelligence Plan, *Marine Corps Vision and Strategy 2025*, *Marine Corps Operating Concepts*, the *MCISR-E Roadmap*, and insights gained from project

---

[4] The campaign plan is a means of promulgating Commandant-level guidance to support ongoing operations while preparing the USMC for the future. There are links between the campaign plan and *Marine Corps Vision and Strategy 2025*. It addresses how campaign plans will be linked to programmatic requirements and decisionmaking processes throughout the USMC.

[5] U.S. Marine Corps Intelligence Department, 2010.

interviews. Starting with a long list of strategic elements, we sought to sort them to connect related items. We then synthesized and combined items until all the derived elements were represented in a relatively parsimonious form. After several iterations and alternative combinations, we found that the seven principles of the 1994 Intelligence Plan provided a good basis for framing current strategic intent. By expanding on those seven principles, we were able to exhaustively represent all USMC intelligence–related elements of strategic intent derived from the aforementioned sources.

While the mission and principles of the original intelligence plan were focused primarily on the operational and tactical levels of the USMC, these seven objectives are meant to aid in assessing and understanding intelligence organization at all levels, so they are necessarily broader. Each of the objectives directly corresponds to one of the seven principles in the 1994 Intelligence Plan but has been broadened, expanded, or tweaked to encompass all the elements of strategic intent that we identified:

- ability to operate in a complex and rapidly changing tactical environment, as well as in complex but more predictable environments
- intelligence capabilities that support decentralized decisionmaking, where appropriate
- intelligence capabilities that are integrated with operations and continuity efforts
- intelligence organizations that are directed by trained, experienced intelligence officers
- intelligence personnel who are trained and practiced in their specialty
- timely intelligence products that are unit- and mission- (not discipline-) focused
- requirements that are understood in the user's terms and context, along with intelligence that is presented and marketed in an accessible way.

Effectiveness and Efficiency

To relate these objectives to organizational design, we first tied them to two broad goals: effectiveness and efficiency.[6] By *effectiveness*, we mean producing useful intelligence; by *efficiency*, we mean doing it with fewer resources. The *MCISR-E Roadmap* recognizes these contrasts as it explains how the roadmap nests within the core competencies of the USMC Service Campaign Plan:

> *Challenges.* Optimization of the MCISR-E to maximize efficiency of intelligence operations must be sacrificed to achieve necessary flexibility and adaptability. Despite this, the need for greater efficiency will not go away, especially with limited intelligence resources available.

> *Mitigation.* Determine what intelligence functions, processes, and supporting infrastructure is directly required to support this competency and design it for the supplest flexibility, accepting the loss in efficiency. But for those functions, processes, and supporting infrastructure not directly supporting this competency[7] but providing foundation baseline intelligence for it, a greater degree of optimization at the price of flexibility and adaptability can be afforded. Articulating where that "dividing line" is must be articulated in detail within the MCISR-E CONOP and be amenable to adjustment over time.[8]

These global goals are not at opposite ends of one scale but form two separate scales. An organization can be more or less effective and more or less efficient to varying respective degrees. Some levels of the overall enterprise should tilt more toward one or the other, as suggested

---

[6] Objectives typically exist in a hierarchy that has, at its apex, the goals of effectiveness and efficiency. Below that level are subordinate objectives that tie to these two global goals. A third level has subordinate statements that help to explain the second-tier objectives. Objectives need to meet their own set of criteria: complete, independent, operational, and small in number.

[7] This refers to USMC core competency 2, which is to employ integrated combined arms across the range of military operations and operate as part of a joint or multinational force.

[8] U.S. Marine Corps Intelligence Department, 2010.

earlier. Our assessment is that in the supporting establishment and CE, efficiency is the dominant goal, while in the combat elements, effectiveness dominates. Obviously, these are broad statements that need to be revisited for each of the subordinate elements in the supporting establishment and CE.

The remainder of this chapter is devoted to a detailed discussion of each of these seven objectives and their connection to other organizational design characteristics (introduced in Chapter Four). We also highlight the relationship of each objective to the goals of efficiency and effectiveness.

Ability to operate in a complex and rapidly changing tactical environment, as well as in complex but more predictable environments. This is an effectiveness objective. The environment is everything outside the boundary of the organizational unit of analysis. As discussed in Chapter Four, Burton et al.'s *Organizational Design* uses two dimensions to describe environment—complexity and unpredictability.[9] Complexity corresponds to the "number of factors in an organization's environment and their interdependency" (p. 48). Unpredictability is a "lack of understanding or ignorance of the environment in terms of the nature of the factors and their variance" (p. 41). Many factors characterize the supporting establishment's intelligence environment, and they are generally well known. This constitutes a "varied" environment. "The varied environment is complex as there are many factors to take into consideration and they can be interdependent (i.e., they influence one another), but these factors are relatively predictable and/or they tend to change within known limits" (p. 45). As one moves from the supporting establishment to the CE and then to the ground, air, and logistics combat elements (GCE, ACE, and LCE), the complexity remains but the unpredictability increases. For the CE, a varied environment still holds. For the combat elements, the environment is best described as turbulent. *Marine Corps Vision and Strategy 2025*, in

---

[9]   Although Appendix A lists a number of references for organizational design, we generally use Burton, DeSanctis, and Obel's *Organizational Design* (2006), as the basis for our assessments.

particular, states that the operational environment is judged to be more complex and the future inherently unpredictable.[10]

Environment matters because, as the literature suggests, organizations with efficiency goals in a varied environment have the best fit by using more hierarchal and functional structures. Organizations with an effectiveness goal in a turbulent environment (complex and unpredictable) have better fit by using flatter, divisional (independent decisionmaking) structures.

In our assessment of issues and courses of action with regard to this objective (see Chapter Seven), we describe the environment for each organizational level discussed earlier. Intelligence organizations must be agile enough to drive (and succeed in) a range of operations. Specifically, they need the ability to prosecute "small wars" and other low-end-of-spectrum contingencies and the ability to assure littoral access and prosecute actions high on the scale of conflict intensity against peer or near-peer adversaries. *Marine Corps Vision and Strategy 2025* states that expeditionary excellence requires transitioning seamlessly among various tasks or performing them all simultaneously.

Intelligence that supports decentralized decisionmaking, where appropriate. This is an effectiveness objective. There are two dimensions that provide insight to whether decentralized decisionmaking fits an organization's environment and structure. The first is whether orientation toward the product or the customer is high or low. A high customer orientation requires a focus on outputs: what the organization is providing to the customer. The outputs or products are sufficiently differentiated to make them important to the customer, indicating a focus on outputs. The second dimension is whether the orientation toward functional specialization is high or low. A high functional orientation requires a focus on inputs, which represent costs to the organization. To the customer, the product is a commodity, and cost matters more than performance.

As stated earlier, organizational units with a high level of functional specialization and a low level of orientation toward product fit best with a functional configuration and a tall or hierarchal structure.

---

[10] Commandant of the Marine Corps, 2008.

In such a configuration, work effort is broken down by department and subunit, with coordination achieved hierarchically through rules and directives. Workflow is from one subunit to another and eventually up through the hierarchy to the executive who allocates resources, makes decisions, and ensures coordination. In the USMC, the Intelligence Department, as part of the supporting establishment, generally takes on a functional configuration, even though some of its subunits may have more of a customer or product orientation (e.g., the Intelligence Estimates Branch, which supports the Commandant of the Marine Corps). But for the most part, highly coordinated, functionally oriented effort is desired.

The combat elements have a divisional configuration. The focus is downward. There is an executive (e.g., division commander or regimental commander) who oversees subunits that are relatively independent of one another. Each of these subunits is externally focused and has its own mission or geographic responsibilities. Because the subunits are relatively autonomous, they can make decisions on their own and meet mission needs in creative ways. Obviously, this configuration describes infantry battalions and regiments but also describes the intelligence subunits that operate at this level. They should have a product and customer organization, and it is clear that, on some level, they do. For example, if the S2A or the intelligence sergeant self-identifies as a 1/1 marine (1st Battalion, 1st Marine Regiment), he or she clearly has a customer orientation. Self-identifying as a SIGINT or HUMINT marine indicates a greater focus on function.

The *Marine Corps Operating Concepts* describes this scenario as command by influence: "decentralization of command with empowered subordinates exercising initiative in accord with the superior commander's intent" (p. 17). *Marine Corps Vision and Strategy 2025* seeks to improve the operational and tactical synergy of the MAGTF so it can better operate in a decentralized manner. In particular, "Tactics, techniques and procedures for disseminating high-value, actionable intelligence down to the lowest tactical level in support of operational maneuver and precision engagements must be further refined" (p. 20). Command-and-control (C2) and intelligence, surveillance, and reconnaissance (ISR) capabilities will be integrated to the squad level to increase the shared situational awareness of small-unit leaders and to support decentralized decisionmaking:

> In environments where human intelligence and tactical information reign supreme, we must acquire and convey information rapidly and accurately to facilitate timely decisionmaking. Over the past decade, we have made great strides in enhancing C2 and ISR at the battalion/squadron level and above. We need now to make similar strides from the battalion down to the squad.[11]

Intelligence that is integrated with operations and continuity efforts. This is an effectiveness objective, based in part on Marine Corps doctrine.[12] *Marine Corps Operating Concepts* states that the USMC may perform a variety of missions across the range of military operations, but two stand at the forefront: assuring littoral access and fighting small wars. What these two operations have in common is that both require forces that are strategically mobile, operationally flexible, and tactically proficient. Moreover, *Marine Corps Vision and Strategy 2025* stresses integration of air and ground-based capabilities for a range of missions. MAGTFs will be optimized to operate as an "integrated system."[13] The command element, including the MEU and the MEB, will be properly equipped with C2, intelligence, communication, and networking systems. "With continuity" is a systems concept that states that subcomponents need to be continuously integrated to remain effective. Thus, intelligence personnel need to be habituated to working with the same tactical units to gain two-way understanding of demand and supply. *Vision and Strategy 2025* promotes the alignment of progressive predeployment training cycles for all MAGTF elements as early as possible in order to build cohesive teams.

---

[11] Commandant of the Marine Corps, 2008, p. 20.

[12] Headquarters, U.S. Marine Corps, *Intelligence Operations*, Washington, D.C., Marine Corps Warfighting Publication 2-1, September 10, 2003, states that the concept for intelligence support "must integrate intelligence activities with operations to provide key intelligence to commanders to enable rapid and effective decisionmaking" (p. 4-11).

[13] Commandant of the Marine Corps, 2008, p. 18.

Chapter Six identifies issues that limit Marine Corps intelligence from achieving these attributes.

Intelligence organizations that are directed by trained, experienced intelligence officers. This is an efficiency objective. It was one of the significant changes in the 1994 Intelligence Plan, and much has been accomplished toward this end in the intervening years. However, given the growth in the number of USMC intelligence officers in the past ten years, the officer force is still relatively junior compared with other areas, such as infantry and artillery. As intelligence officers "age" in the closed officer personnel system, their grade distribution (O-1 to O-6) will begin to mirror that of their peers. However, traditional operational and command assignments are more difficult to attain for intelligence officers. As a result, overall career patterns will not mirror those of many of their peers, and promotion boards will need to understand these differences. The USMC must either provide command opportunities to intelligence officers so that their career pattern looks like that of other officer career fields or establish mechanisms to mitigate the effect. Each service uses slightly different procedures for promotion competition for intelligence officers.[14] Furthermore, each service uses promotion board precepts from the service secretary to illustrate special considerations for some groups. *Marine Corps Operating Concepts* states that a mix of incentives and specific precepts for promotion opportunities may be needed for adequate career management. Intelligence officers must be able to do intelligence while having viable career paths. *Marine Corps Vision and Strategy 2025* promotes professional military education as a career-long activity that must be increasingly focused on junior marines. Some of the issues raised in

---

[14] In the Army, Military Intelligence (a branch) and Strategic Intelligence (a functional area) are parts of the Operations Support competitive category. Also in that category are signals, foreign area officers, space, academy professors, operations research and systems analysts, force management, nukes, strategic operations, and strategic plans and policy. In the Navy, intelligence is a separate special-duty officer competitive category. In the Air Force, intelligence is included in the line of the Air Force competitive category, an arrangement that is most similar to the Marine Corps. So, promotion competition for intelligence careers spans the spectrum from its own competitive category to competing with everyone but the professionals (e.g., physicians, lawyers).

Chapter Six relate to providing more or fewer opportunities for command and critical assignments.

Intelligence personnel who are trained and practiced in their specialty. This is an efficiency objective. In the 1994 Intelligence Plan, one of the focal "functional concepts" was that "staffs use, organizations produce" intelligence. According to the plan, this concept was a response to concerns about the training and readiness of intelligence marines. Even now, one of the DIRINT's focus areas is to professionalize the intelligence workforce and ensure that personnel possess the proper aptitude, training, education, and experience (advanced technical skills and regional/cultural expertise) to understand the complex hybrid threat and provide commanders with the necessary critical analysis to make sound, timely decisions. Analysis, as a skill, is particularly important, and proficient analysts are the product of both selection (for inherent characteristics and abilities) and development (of attributes, knowledge, and skills). *Marine Corps Operating Concepts* suggests that assignment patterns and policies should be aligned with support regionalization through repeated tours in units that concentrate on specific regions. *Marine Corps Vision and Strategy 2025* states that training must be tailored to develop cohesive units. Training and education must accurately reflect the situations, environments, and people marines will face. Intelligence personnel must be sufficiently exposed to tactical needs and operational concerns to understand the needs of the combat elements. We describe issues in Chapter Six that can affect the ability to train and practice.

Timely intelligence products that are unit- and mission- (not discipline-) focused. This is an effectiveness objective. *Marine Corps Operating Concepts* outlines capabilities that are needed to conduct operations by organizational level, as shown in Table 5.1.

Marine Corps Vision and Strategy 2025 states that robust intelligence capabilities will support all levels of command awareness and decisionmaking. Moreover, integration of ISR across the MAGTF has to be more aggressively explored. The *MCISR-E Roadmap* has as a focus area "integration of all Service ISR elements into a holistic system, networked across all echelons and functions." Some of the issues identified

Table 5.1
Needed Intelligence Capabilities

Organizational Level	Intelligence Capability
Platoon	Data feed from company-level intelligence cell (CLIC)
Company	HUMINT, MASINT cell, unmanned aircraft system (UAS) access (fed from MAGTF), connection between CLIC and MAGTF S2
Battalion	Human terrain/environmental, target populations
MEU	Human terrain/environmental, target populations
MEB	Human terrain/environmental, target populations
MCIA	Service intelligence production center; provides reachback for tailored expeditionary intelligence analysis and cultural studies; provides highly focused predeployment training; augments units with specialized teams and liaison officers

SOURCE: Deputy Commandant for Combat Development and Integration, 2010, pp. 46–48, 61.

in Chapter Six affect the timeliness and mission orientation of intelligence for these units.

Requirements that are understood in the user's terms and context, along with intelligence that is presented and marketed in an accessible way. This is an effectiveness objective. *Marine Corps Operating Concepts* lists intelligence as a warfighting function. Collection and dissemination enhancements provide the maneuver forces and the MAGTF commander with greater insight into the enemy and the context of the battlespace: Intelligence and information should "flow throughout the force in a rapid, palatable manner" (p. 43). Under the rubric of enhanced MAGTF operations, the document states that the essence of intelligence is the ability to process information and knowledge at the point of action; at its crux are real-time collection, fusion, intuitive products, and dissemination of those products; and enhancements are needed to establish intelligence cells at the lower levels, improve intelligence networks, and tailor, automate, and balance information pull (passive) and selective push. Intelligence capabilities must be consistently exposed to commanders and staff at all levels who can provide necessary understanding of the capabilities and use, not just dissemination, of USMC intelligence products.

Issue Identification and Analysis of Data

This chapter describes how we identified issues of concern to USMC intelligence through our interviews with a range of USMC personnel and civilians and how we prioritized those issues based on the seven objectives articulated in Chapter Five.

As discussed in Chapter Two, our qualitative data analysis of the interview transcripts using ATLAS.ti led us to identify and categorize 48 unresolved issues related to USMC intelligence. Each issue was raised by multiple respondents; details and examples, as well as specific comments from the interviews, are presented later in this chapter.

Ranking the Issues

To provide some structure to our analysis, we began by categorizing the 48 identified USMC intelligence issues. Although our interviews focused on USMC intelligence *organizational* topics, some of the concerns and issues raised were not explicitly organizational or structural. We retained issues that were not strictly organizational because their identification may still be useful to the organization. We sorted the identified issues into two categories: structural and not structural. Note that these categories do not contribute to or constrain the analysis in any way; we use them solely for ease and clarity of presentation.

We also sought to prioritize the issues. We elected to do so based on each issue's possible impact, if left unresolved, on the seven objectives identified in the previous chapter. To this end, we first considered each issue against each objective and then scored the issue as

(1) a challenge that threatens the objective, (2) a risk factor that could adversely affect the meeting of the objective, (3) not adversely related to the objective and unlikely to become so even if conditions change, (4) not adversely related to the objective but at risk of becoming so if conditions change (trade-offs), or (5) not applicable, or unrelated to the objective. We then gave each of the five conditions a quantitative score: 1 = 0.5; 2 = 0.25; 3 = 0.05; 4 = 0.2; 5 = 0. These are ratio scores of the risk that the issue poses to a given objective.

We then assigned weights to each of the objectives identified in Chapter Five (for convenience, they are presented in priority order in that chapter).[1] To complete the prioritization process, we took the ratio score for each issue-objective set, multiplied it by the weight for the objective, and then summed across objectives.[2] This resulted in an overall score of the possible threat or risk for each identified issue. For example, the issue of "intelligence personnel doing collateral duties instead of intelligence" directly challenged the six most highly ranked

---

[1] We did this by overweighting higher-priority objectives and underweighting lower-priority objectives using calculated weights based on rank-ordering the objectives from 1 to 7. Analytically, the method drives greater separation among issues. See F. Hutton Barron and Bruce E. Barrett, "Decision Quality Using Ranked Attribute Weights," *Management Science*, Vol. 42, No. 11, November 1996, for an empirical assessment of the role of such formulas in significantly improving decision quality. Specifically, our rank ordering of the objectives was (weight in parenthesis): (1) intelligence capabilities that are integrated with operations and continuity efforts (0.37); (2) timely intelligence product that is unit- and mission- (not discipline-) focused (0.23); (3) intelligence capabilities that support decentralized decision-making, where appropriate (0.16); (4) requirements that are understood in user's terms and context, along with intelligence presented and marketed in an accessible way (0.11); (5) intelligence personnel who are trained and practiced in their specialty (0.07); (6) ability to operate in a complex and rapidly changing tactical environment, as well as in complex but more predictable environments (0.04); (7) intelligence organizations that are directed by trained, experienced intelligence officers (0.02). Weights across all seven objectives sum to 1.

[2] We then multiplied these threat values by 2, so that any issue that challenged or threatened all objectives (score of 0.5 on each objective), when multiplied by the weights for each objective (which sum to 1), would yield a weighted threat value of 1. Threat values thus hypothetically range from 0 (issue is not applicable or is unrelated to all objectives) to 1 (issue is a challenge or threat to every objective), with intermediate values reflecting the number of objectives challenged, the extent of the threat, and the weight or importance of the objective threatened.

of the seven objectives and rose to the top of the list as a result.[3] The issue "need for command billets" could adversely affect one of the lower-ranked objectives and was a risk to another lower-ranked objective if conditions changed; thus, its score was low overall. The highest scores were assigned to issues that affected objectives with the highest priority.[4]

Table 6.1 presents the 48 issues in priority order, along with their priority scores. The table also notes whether the issue is primarily structural/organizational or not and lists the principal organizational levels to which the issue applies.

In the next section, we provide details on the issues as raised in the interviews.

Why Include Nonstructural Issues, and Why Prioritize?

Given the explicit focus of this monograph on organizational structure, there is room to question why we have chosen to present so many issues that are either tangential or wholly unrelated to structure and why we have attempted to rank or prioritize this collection of issues by importance. Certainly, we could have used our extensive set of interviews solely as a holistic source of problems to be addressed through structural change. Indeed, the interviews served admirably for that purpose.

---

[3] So, "intelligence personnel doing collateral duties instead of intelligence" scored [(0.5 × 0.37) + (0.5 × 0.23) + (0.5 × 0.16) + (0.5 × 0.11) + (0.5 × 0.07) + (0.5 × 0.04) + (0 × 0.02)] × 2 = 0.98, the highest threat/risk score observed.

[4] So, "need for command billets" was a risk factor (0.25) for the objective that requirements be understood in the user's terms and context and that intelligence be presented and marketed in an accessible way (the fourth-ranked objective). It is not adversely related—but at risk of becoming so if conditions change (0.2)—to the objective that intelligence organizations be directed by trained, experienced intelligence officers (the seventh-ranked objective), resulting in a threat/risk score of [(0 × 0.37) + (0 × 0.23) + (0 × 0.16) + (0.25 × 0.11) + (0 × 0.07) + (0 × 0.04) + (0.2 × 0.02)] × 2 = 0.063, the lowest threat/risk score observed. Note that we did conduct some sensitivity analysis of the prioritizations by varying the weights assigned to the objectives. Varying the weights had a minimal impact on the priorities assigned. In practice, issues that were risks or possible risks to multiple objectives rose to the top under any weighting scheme.

Table 6.1
Prioritized Issues

Priority Rank	Issue	Threat/Risk Score	Category	Organization
1	Intelligence personnel doing collateral duties instead of intelligence	0.98	Nonstructural	CE mainly, but MEF and intelligence battalion too
2	Vicious cycle in aviation: intelligence not well prepared to support aviators; aviators view intelligence as irrelevant	0.96	Nonstructural	ACE
3	General propensity to respond to the commanding general's curiosity rather than tactical force needs	0.925	Nonstructural	MEF, intelligence battalion
4	MCIA lacks a 24-hour watch cycle	0.905	Structural	MCIA
5	Regurgitation, not analysis, from certain layers in the organization	0.9	Nonstructural	MEF, intelligence battalion, CE
6	Misuse of reconnaissance or intelligence assets by controlling maneuver forces	0.9	Nonstructural	CE
7	Stovepiped intelligence specialty "tribes"—lack of integration or desire to build single intelligence-only targeting profiles	0.891	Structural	I-Dept, MCIA, MEF, intelligence battalion, radio battalion, CE
8	Layers of bureaucracy/layers of intelligence; no clear guidance on what is to be done at different levels	0.89	Structural	MEF, intelligence battalion, CE
9	Intelligence officers who do not understand tactical context and thus provide limited value	0.89	Nonstructural	MEF, intelligence battalion, radio battalion, CE

Table 6.1—Continued

Priority Rank	Issue	Threat/Risk Score	Category	Organization
10	Priority intelligence requirements (PIRs) are not updated often enough	0.845	Nonstructural	MEF, CE
11	In current operations, the majority of manpower is in the intelligence battalion, but the majority of information is in the infantry battalion	0.83	Structural	MEF, intelligence battalion
12	Issues with MCIA websites	0.815	Nonstructural	MCIA
13	Collection requirements management is neglected	0.81	Nonstructural	MEF, intelligence battalion, radio battalion
14	Seniority and experience needed at the lowest tactical level	0.785	Structural	CE
15	Flexibility needed to push personnel out to the tactical level and pull them back if needs change or if they are being misused	0.775	Structural	MEF, intelligence battalion, radio battalion
16	Authority conflated between intelligence battalion commanding officer and MEF G-2; subordinates can be pulled in different directions, and priority follows the fitness report	0.748	Structural	MEF, intelligence battalion
17	Lack of integration, coordination, and focus—I-Dept very stovepiped	0.745	Structural	I-Dept
18	MCIA serves multiple masters: USMC Headquarters, operating forces, DIRINT, IC, MCCDC; struggles to say no; lacks boundaries in terms of mission and lacks clear mission statement	0.74	Structural	MCIA

Table 6.1—Continued

Priority Rank	Issue	Threat/Risk Score	Category	Organization
19	Some tactical commanders do not understand intelligence	0.71	Nonstructural	CE
20	In garrison, the intelligence battalion does minimal intelligence work	0.705	Nonstructural	Intelligence battalion
21	Personality-driven command relationships	0.665	Nonstructural	MEF, intelligence battalion
22	I-Dept consumed with treading water; struggles to look past immediate firefighting to long-term vision	0.65	Nonstructural	I-Dept
23	Impositions and informal tasking from I-Dept to MCIA	0.65	Structural	MCIA
24	DIRINT has limited direct authority	0.63	Nonstructural	I-Dept
25	The intelligence battalion trains as a battalion but does not deploy as a battalion	0.616	Structural	Intelligence battalion
26	Need to man for increased collection assets in ACE (i.e., Joint Strike Fighter [JSF], UAS)	0.59	Structural	ACE
27	SIGINT personnel useless without their tools and database access	0.585	Nonstructural	Radio battalion, CE
28	I-Dept legacy branch and division alignment/branches misnamed rather than misorganized	0.558	Structural	I-Dept
29	I-Dept undermanned for full extent of tasks and short on department/branch heads	0.557	Structural	I-Dept

Table 6.1—Continued

Priority Rank	Issue	Threat/Risk Score	Category	Organization
30	Many different views of MEF Intelligence Center (MIC); confusion	0.545	Nonstructural	Multiple organizations
31	Concerns about analysts—difficult to "make" an analyst (something intrinsic and a lot of training needed)	0.525	Nonstructural	Multiple organizations
32	Bandwidth issues at the tactical level on certain deployments and a lack of connectivity in garrison	0.525	Nonstructural	Intelligence battalion, CE
33	Concerns about I-Dept and DIRINT grade structure relative to the rest of the USMC and the broader IC	0.52	Nonstructural	I-Dept
34	All USMC intelligence officers go from specialist to generalist	0.501	Nonstructural	Multiple organizations
35	Difficulty matching correct assignment (general support, direct support, other) with detachments; "multiple masters"/problems with general support	0.49	Structural	MEF, intelligence battalion, radio battalion, CE
36	I-Dept lacks a single entry point and advocate for GCE, LCE, and ACE	0.49	Structural	I-Dept
37	Need seniority and experience to conduct quality, high-level analysis and represent intelligence within and outside the USMC	0.472	Nonstructural	I-Dept, MCIA, MEF, intelligence battalion, radio battalion
38	Enterprise-wide issues with knowledge management and information management	0.445	Nonstructural	Multiple organizations
39	Lots of informal networking (buddy network) required to get things done	0.41	Nonstructural	Multiple organizations

Table 6.1—Continued

Priority Rank	Issue	Threat/Risk Score	Category	Organization
40	Limited I-Dept budget	0.365	Structural	I-Dept
41	Grade and seniority issues relative to others in the USMC and at the interagency level (e.g., 1-star in a 3-star world; X2 always junior to X3)	0.33	Nonstructural	I-Dept, MCIA
42	Concerns about lateral entrants—insufficient screening/requirements, placement in "experienced" positions without experience	0.33	Nonstructural	Multiple organizations
43	Excessive bureaucracy in MCIA, with command and subordinate commands	0.303	Structural	MCIA
44	In an intelligence battalion, almost no intelligence officers do intelligence work; most are consumed by administrative, management, oversight, and command tasks	0.285	Structural	Intelligence battalion
45	Struggle to capture or share new solutions or procedures, which leads to reinventing the wheel (includes lack of tools for capturing and sharing lessons learned)	0.145	Nonstructural	Multiple organizations
46	Command experience required for promotion	0.135	Nonstructural	Multiple organizations
47	Standards and tools to contribute to national intelligence not in place	0.115	Nonstructural	Multiple organizations
48	Need for command billets	0.063	Nonstructural	Multiple organizations

However, they also provided a rich collection of issues and challenges that should be of interest to those who wish to continue to improve the USMC intelligence enterprise. We have chosen to present all of these issues, whether they are organizational or not and whether they might be addressed through structural change or realignment, because addressing them through other means could still provide benefit to the USMC.

Since our interviews were not conducted with a statistically representative sampling frame, we do not (and should not) report the frequency with which each issue was raised, and such numbers would be an inappropriate means by which to assign importance to these issues. If the USMC seeks to address some of these issues, it would benefit from some idea of which are most important or may have the greatest impact. With that end in mind, we developed and employed the ranking scheme described earlier in this chapter.

Issues

In this section, we elaborate on each of the 48 issues listed in Table 6.1. Note that they remain in priority order. Each issue is preceded by its priority rank out of 48.

1. Intelligence personnel doing collateral duties instead of intelligence. Many respondents reported and were concerned about intelligence marines being tasked with collateral duties to the exclusion of intelligence tasks, both while deployed and in garrison. None suggested that intelligence marines are not marines first or that they should be excused from collateral duties. Several respondents, however, reported anecdotes or personnel experiences that involved intelligence marines being kept from doing their intelligence jobs by excessive collateral burdens.

The overall assessment was that intelligence marines are either underutilized or misused and that they have greater capabilities than recognized or given credit for. Too often, they are tasked with administrative duties instead of actual intelligence work.

The collateral duties of concern range from having sole responsibility for handling clearances within a unit (since intelligence marines are both sufficiently cleared and have access to classified networks) to "painting rocks" (unfavorably viewed garrison administrative duties). In the words of one respondent, "They didn't allow us to be intel marines; they don't know what intel can do for them."

2. Vicious cycle in aviation: intelligence not well prepared to support aviators; aviators view intelligence as irrelevant. Several respondents reported issues with intelligence support for USMC aviation. The bottom-line consensus appears to be that intelligence officers are not pilots and thus face an uphill struggle for credibility and perceived value in the aviation community. For example,

> Young aviation intel officers are not well prepared for or respected in the wings. They are thrown out to staff jobs too early in their career. They are given no way to relate to the pilots; they speak different languages.

Compounding the difficulty of this cultural tension between intelligence personnel and aviators is the observation that "intel is viewed as irrelevant in the aviation community." What starts as a cultural mismatch then festers and gets worse:

> It becomes a vicious cycle. Lack of respect for a young, underinformed aviation intel officer shows in mission reports, which leads to aviators blowing off debriefs, which impairs the intel officer's ability to do the job, which prevents him from getting the information needed to paint the broader picture and show value.

This cycle is perpetuated by the fact that many aviation intelligence officers have bad experiences and leave the corps or the wing. Their replacements are often other young and underprepared aviation intelligence officers who do nothing to change aviators' negative views of intelligence personnel, according to our interviewees.

3. General propensity to respond to the commanding general's curiosity rather than tactical force needs. Many respondents decried a general propensity to answer the commanding general's curiosity rather than meet the intelligence needs of tactical forces. They described significant intelligence formations (for example, the MEF G-2 and the intelligence battalion preoccupied with "feeding the bear"). According to one respondent,

> Intel and radio battalions have a reputation as a self-licking ice cream cone. They produce for MEF and the [commanding general] and not for subordinate commands.

Another noted that

> too much manpower is devoted to providing big products for the [commanding general], especially the Graphic Intelligence Summary.

> [This arrangement is] unfortunate since requirements at the tactical level are often drastically different than requirements at the command level.

Many respondents view this propensity as particularly harmful, especially in a COIN context. As one respondent observed,

> It doesn't matter what the MEF commander sees in his slides. What is important is that the 19-year-old ready to go on patrol knows what he might find.

4. MCIA lacks a 24-hour watch cycle. Currently, MCIA does not have personnel on duty for 24 hours a day, nor is it able to stand up a 24-hour watch. As one interview respondent noted, this is "particularly rough on swing time zones, like much of [U.S. Pacific Command]." The lack of a 24-hour watch is compounded by concerns about MCIA's web page. One respondent summed up the concern with this comment: "MCIA is great, if they pick up the phone." Implied, however, was that when MCIA does not pick up the phone—whether because it is after business hours at Quantico or for some other reason—the marine making the call lacks confidence in MCIA's ability to respond to requests or provide data.

Several respondents suggested that if MCIA is to retain or expand its reachback role, it should go to a 24-hour watch structure. When reachback needs are considered urgent or require close collaboration for rapid iteration, it can be impossible to meet those needs when the individuals collaborating are on asynchronous watch cycles.

5. Regurgitation, not analysis, from certain layers in the organization. Numerous respondents expressed frustration about their experiences with different layers of the USMC intelligence structure during deployments in Iraq or Afghanistan. For example, one respondent complained that during his rotation in Iraq, the

> tactical fusion centers provided little value added with their [intelligence summaries] and their intel products. The intel products were often regurgitations that were three days old. These centers typically were not getting enough information, so their products were sometimes incomplete and written out of context.

According to another respondent,

> The fusion cell didn't get good enough information from the tactical level to do good work. They were 100-percent reliant on the G-2s and S2s to push information up to them. Ended up being mainly PowerPoints made from others' PowerPoints.

This concern was not unique to tactical fusion cells; it was reportedly a problem at all levels whenever synthesis was attempted based on finished products or briefings rather than broader (or more raw) information and data. This issue is not unique to our interviews. It also received attention in the *Marine Corps Gazette*.[5]

6. Misuse of reconnaissance or intelligence assets by controlling maneuver forces. Another issue raised by several respondents that relates to leadership more than organizational structure concerned the misuse of intelligence or reconnaissance assets. This extends beyond the collateral duties issue raised earlier (issue 1). The most egregious

---

[5] See, for example, Michael P. Foley, "Facilitating Intelligence at the Point of Action," *Marine Corps Gazette*, Vol. 94, No. 3, March 2010.

example was an anecdote recounting unmanned aerial vehicles (UAVs) being used to watch a convoy's progress rather than to reconnoiter the convoy route or watch for threats approaching or lying in wait for the convoy.

This issue was not unique to our interviews. A 2007 *Marine Corps Gazette* article noted that, in the author's experience in OIF,

> Instead of being dedicated to intelligence requirement–driven collections, imagery and UAS assets are predominantly used to conduct continuous raid coverage, providing live video for the commander and satisfying his desire for dynamic information.[6]

7. Stovepiped intelligence specialty "tribes"—lack of integration or desire to build single intelligence-only targeting profiles. The stovepipes of the intelligence specialty "tribes" are nothing short of legendary, and also a concern outside the USMC.[7] Reportedly, functional stovepipes permeate many organizations and levels of USMC intelligence organizations. USMC Systems Command, I-Dept, intelligence battalions, and radio battalions were all specifically mentioned as organizations in which intelligence-based functional stovepipes result in dysfunction. SIGINT and HUMINT were the areas most frequently described as having their own tribe or exclusive stovepipe. While many respondents reported successful all-source integration, several pointed to specific instances of stovepiping that led to adverse operational results, such as one officer's anecdote about SIGINT personnel who preferred to generate "SIGINT-only" targeting profiles that proved to be deficient relative to all-source, synergized products.

8. Layers of bureaucracy/layers of intelligence; no clear guidance on what is to be done at different levels. Similar to the observation that certain layers in the organization are more prone to regurgita-

---

[6] Jeffrey Dinsmore, "Intelligence Support to Counterinsurgency Operations: The Search for Fused, Coherent Intelligence to Support the Commander," *Marine Corps Gazette*, Vol. 91, No. 7, July 2007, p. 25.

[7] See, for example, the attack on INT stovepipes in Michael T. Flynn, Matt Pottinger, and Paul D. Batchelor, *Fixing Intel: A Blueprint for Making Intelligence Relevant in Afghanistan*, Washington, D.C.: Center for a New American Security, January 4, 2010a.

tion than analysis, respondents opined that some levels in the deployed intelligence hierarchy did not make a useful contribution because of a lack of clear guidance about the types contributions expected from different levels. To wit: There are "no clear assignments of what should be added or done at each level of the hierarchy."

Others raised concerns about the lack of clear focus elsewhere in the USMC intelligence enterprise. One respondent suggested that "one problem with MCIA is that they do not have a clear and defined mission statement," which adversely affects its ability to prioritize (see also issue 4).

9. Intelligence officers who do not understand tactical context and thus provide limited value. A recurring theme in many of the interviews was the imperative that intelligence officers understand the tactical context of the forces for which they are providing intelligence support, be they GCE, ACE, or LCE. Intelligence personnel who do not understand this context (for whatever reason, e.g., too junior, lacking context or specific experience) will provide limited value.

> The intel guys need to have an understanding of infantry, aviation, logistics tactics, and what the combat element guys need so they can provide it without too much guidance. You need to give me exactly what I need, and you need to know what I need.

Several respondents opined that ground intelligence officers offer very good tactical understanding to the GCE: "A ground intel officer knows everything that his counterpart in the infantry battalion does, and intel, too." Most often, the issue was thought to apply to lateral entrants or intelligence personnel assigned to the ACE.

10. PIRs are not updated often enough. Several respondents noted that PIRs are not updated with sufficient frequency. One respondent indicated that, in some operational areas, they were updated only annually or semiannually, while they should have been revisited much more frequently. "Official requirements (priority intelligence requirements, intelligence requirements) aren't always updated to reflect new operational objectives."

Stale PIRs impair intelligence efforts, because "PIRs and IRs: That is how you task intel."

11. In current operations, the majority of manpower is in the intelligence battalion, but the majority of information is in the infantry battalion. Many respondents specifically confirmed the received wisdom that, in COIN operations or other "small-war" activities, intelligence needs to be pushed down to the lowest tactical level. Regarding contemporary COIN operations, one respondent noted that the "majority of manpower is in the intel battalion, but the majority of information is at the infantry battalion. Need to turn this around."

Respondents suggested that COIN operations require not only greater numbers of intelligence personnel at lower levels but also greater intelligence expertise, experience, and capabilities among those personnel. Of course, the need for intelligence personnel at the lowest level may be exclusive to small-war situations. Other types of operations may require that intelligence experience and expertise be distributed elsewhere; many respondents acknowledged this tension.

12. Issues with MCIA websites. Several respondents mentioned problems with MCIA websites. Here are some examples:

> Their web page is not designed for easy access. Have to pick up the phone and call.

> MCIA changed their website; it is now more confusing.

> Because of confusing changes to the MCIA website, we now use the NGA website more.

13. Collection requirements management is neglected. Related to the infrequency with which PIRs and IRs are updated (see issue 10) is the collection requirements management process. According to one respondent,

> Collections management is dysfunctional in the Marine Corps and DoD-wide. At the division level, especially, coordination between different elements is lacking.

Another noted,

> Collections also suffers from lack of tasking discipline. While
> he was deployed, the G-2 took some ISR sortie time away from
> some of the maneuver battalions to force some tasking disci-
> pline on them. This worked and made them come up with more
> focused tasks and requirements for their assets. Once that G-2
> left, though, everyone went back to being sloppy.

When collection requirements are not well managed or requirements are not passed effectively to and from different levels and organizations, the data needed for analysis will not be available. In addition,
some information could be unnecessarily redundant, or requirements
generated elsewhere in the IC might not be met.

14. Seniority and experience needed at the lowest tactical level.
This point echoes issue 11. Respondents noted that intelligence manpower was predominantly in the intelligence battalion and not down
in the infantry battalion. They were also concerned that those with
seniority and experience were not as close to the fight as they needed to
be. One respondent noted,

> Too many senior [noncommissioned officers] and senior captains
> hide themselves in headquarters billets, away from the tactical
> level. They aren't available to share their experience.

Another observed,

> Senior gunnys, others, have some of the best insights. It is a par-
> adox: The more senior you get, [the more] you tend to bubble
> toward the top, rather than being at the lowest level where your
> matured skills can make the most contribution.

Again, respondents noted that a "seniority-forward" arrangement
is particularly important for COIN and small war–type operations,
but it may not be what is needed for operations elsewhere in the range
of military operations.

**15. Flexibility needed to push personnel out to the tactical level
and pull them back if needs change or if they are being misused.**

Further conversation with interview respondents who were concerned about the limited intelligence manpower, experience, and expertise at the lowest tactical level revealed that they anticipated a need for different workforce allocations for operations in different types of military operations. These respondents discussed a need for flexibility and the ability to push expertise out and down when needed, but to contract and centralize that expertise when necessary, too.

Others raised concerns that intelligence personnel pushed out to tactical formations might not be employed optimally by tactical commanders. Current structures in which such personnel are detailed to those formations rather than assigned to them allows the high-level commander the flexibility to retrieve those personnel if they are being misused. Retaining such flexibility was highlighted as a virtue.

16. Authority conflated between intelligence battalion commander and MEF G-2; subordinates can be pulled in different directions, and priority follows the fitness report. Many respondents reported an issue that emerges when the MEF G-2 and the intelligence battalion commander, deployed as the intelligence support coordinator, do not reach a shared understanding and agreement about responsibilities. In such situations, deployed intelligence battalion personnel might receive conflicting orders, tasks, or instructions from both the MEF G-2 and the battalion commander. One respondent said,

> This is rough on the company commanders. An O-6 G-2 is giving orders past the O-5 battalion CO, who, by the way, is the one who write the company commanders' fitreps. It is supposed to be amicable, but there is tension within.

This could be further compounded if intelligence battalion personnel were part of a detachment assigned to a maneuver formation, in which case they might receive direction from the G-2, the intelligence battalion commander, and the commanding officer of an infantry battalion, for example. In the words of one respondent,

> My CO writes my fitrep. The G-2 writes tasks. I'm sent out to the infantry battalion. There needs to be doctrine or policy for who has definitive control [and] when.

Personality plays a major role. For example, with regard to intelligence battalion COs,

> Some of them just want to be the commander but not the intel guy or the intelligence support coordinator, and they want to retain control of "their guys" and be "the commander."

Respondents reported that receiving conflicting instructions from multiple sources of authority is, unsurprisingly, stressful. They stated that, in practice, such tension is always resolved in the direction of the fitness report. Intelligence marines will try to make everyone happy, but if the guidance conflicts, they follow the guidance from whomever is going to write their fitness report.

17. Lack of integration, coordination, and focus—I-Dept very stovepiped. In addition to concerns about functional stovepipes (see issue 7), several respondents raised related issues about I-Dept. One respondent reported stovepipes as alive and well in I-Dept, indicating, "What cross-INT [specialty] leveraging there is, is all based on informal liaison." Respondents observed that odd functional alignments and stovepiping adversely affect integration and coordination at I-Dept. For example,

> In I-Dept, stuff at the [action officer] level, the branch level, is kind of strange. Weird lack of communication between branches. Internal communication and process issues. Good people, bad processes. Dysfunctional.

18. MCIA serves multiple masters: USMC Headquarters, operating forces, DIRINT, IC, MCCDC; struggles to say no; lacks boundaries in terms of mission and lacks clear mission statement. Several respondents suggested that MCIA has an excess of missions and masters. For example, "MCIA answers to about four different masters, and they are torn in a million different directions."

Part of the problem, one respondent opined, "is that they do not have a clear and defined mission statement." An officer who previously served in MCIA candidly admitted,

> We have no methodology to prioritize what to work on. [National Intelligence Priorities Framework], DoD guidance, etc., provide some priorities, but there is not a single, clear list.

A lack of priorities and multiple masters create a frenetic pace of work at MCIA. Since priorities are not clearly identified (one respondent told us that "the operating forces are priority-one" but refused to discriminate beyond that), MCIA personnel cannot say "no" to lower-priority customers.

19. Some tactical commanders do not understand intelligence. Echoing issue 9, that intelligence officers who do not understand the tactical context add limited value, is the related issue that some tactical commanders do not understand, or understand how to use or task, intelligence. Respondents noted that if commanders do not understand the intelligence process (in terms of lead times required), procedures for generating and prioritizing requirements, or simply what is possible, they are less able to request what they really need and more likely to be frustrated with the intelligence they ultimately receive.

20. In garrison, the intelligence battalion does minimal intelligence work. Multiple respondents observed that the intelligence battalions do minimal intelligence work in garrison. They ascribed this to several causes. First, many noted insufficiencies on the technical side: lack of equipment, lack of classified network connectivity, or insufficient numbers of classified computer terminals. Second, respondents highlighted competing activities, including routine and specialized training needs and an excess of collateral duties in garrison ("intelligence marines painting rocks in garrison"). Third, respondents reported a general failure to exercise relationships with the MEF, the higher-level staffs, or frequently supported maneuver units while in garrison:

> In garrison we don't exercise too well our relationship with the MEF or the higher-level staff. When we deploy for a real-world contingency, we do better because we are under the G-2. In garrison, there are no intel requirements from the G-2, but there are personnel requirements.

21. Personality-driven command relationships. A recurring sub-ordinate issue in many of the interviews surrounded personality-driven command relationships. This issue could be resolved or potentially worsened depending on the personalities involved. For example,

> The G-2/G-3 relationship is personality-driven. They either work together or at odds with one another. Most of the time, it's in the middle. They don't always communicate well.

With regard to the hospitality received by detachments in garrison:

> The infantry battalions should provide the care and feeding for the [detachments] so they can do their jobs. Personality over doctrine made things confusing, especially at the bottom.

Many other relationships were described as personality-dependent: the relationship between the MEF G-2 and the intelligence battalions; the relationship between I-Dept and MCCDC or the USMC Training and Education Command (TECOM); informal relationships between elements of I-Dept and elements of MCIA; the relationship between intel battalion and radio battalion commanding officers.

22. I-Dept consumed with treading water, struggles to look past immediate firefighting to long-term vision. Several respondents reported concerns about I-Dept's focus and emphasis, noting that I-Dept is consumed with day-to-day "firefighting." While some of this no doubt stems from workload relative to manning, respondents suggested that it also stems from embracing too broad a mission. One respondent asked rhetorically, "Why are we [I-Dept] doing current intelligence?" and suggested that I-Dept is stretched thin because of broader IC and Title X requirements and its mission to support the Commandant with current intelligence. Another respondent believed that I-Dept is too concerned with what the MARFORs should focus on, but those tasks are too low-level, too "into the weeds" of operations. The respondent thought that I-Dept "should be looking farther forward."

23. Impositions and informal tasking from I-Dept to MCIA. Several respondents reported significant impositions and informal taskings from I-Dept to MCIA. None disputed that MCIA works for I-Dept or that I-Dept has every right to task MCIA. Complaints focused on one of two areas. The first was when I-Dept simply passed along tasks that a respondent believed I-Dept itself should have completed. The second was when individuals in I-Dept contacted subordinate personnel in MCIA directly with a task rather than passing it through the chain of command, which would allow MCIA command personnel and managers to retain awareness of individual personnel assignments and workload. For example,

> Too much informal tasking [from] person to person, ignoring the chain of command. There are no rules, no [standard operating procedures] for sharing information up and down; this makes it personality-dependent. There's no staffing process. There is a formal tasking process, but it isn't well used. This results in [I-Dept] direct-tasking to MCIA staff without the knowledge of the MCIA management.

24. DIRINT has limited direct authority. Several respondents questioned the extent of DIRINT's ability to make changes in the structure and organization of USMC intelligence. One respondent described DIRINT's authority in this way:

> He has specified roles within the expeditionary force order but not a lot of directive authority. More "bully pulpit" influence. He would need institutional support for any big changes, including "energy" from other key flag officers, such as [in] MARFOR or MEF.

According to some respondents, in addition to lacking compelling authority in certain areas of USMC intelligence and needing to rely on persuasion, reason, and influence to realize significant organizational changes, DIRINT should have greater authority over day-to-day interactions with other parts of the enterprise. For example, "DIRINT should have had more authority over the intel integration division, which falls under MCCDC."

25. The intelligence battalion trains as a battalion but does not deploy as a battalion. Several respondents noted something to the effect that the

> intelligence battalion has a different structure in garrison and deployed. In garrison, there are traditional companies and platoons. When deployed, the structure is different. [Direct support teams] and RCTs [regimental combat teams] are pulled from platoons. Some of the platoons stay intact, but others are dissolved into "cells" and to lead [direct support teams].

The concern echoed by numerous respondents was, "We don't train as we fight."

26. Need to man for increased collection assets in ACE (JSF, UAS). Several respondents noted intelligence manpower shortfalls in the ACE, especially with regard to new systems, including JSF and various unmanned UASs. The issue of intelligence manning for JSF is particularly tricky, first, because of (1) the time it takes to change the Table of Organization (T/O) to generate needed billets, (2) the length of training pipelines to generate needed skills, and (3) uncertainty about exactly which collection capabilities will appear on JSF airframes and under what timeline.

27. SIGINT personnel useless without their tools and database access. Several respondents informed us that SIGINT personnel are wholly dependent on their specialized tools and access to NSA databases to make a contribution. Respondents were concerned that this fact is not always appreciated by combat element commanders. For example,

> Infantry battalion commanders don't always realize that if SIGINTers aren't connected to the enterprise, they aren't good for much. They need the [sensitive compartmentalized information] comms to function.

Another noted that "SIGINT is heavily dependent on reachback but very bandwidth-intensive."

28. I-Dept legacy branch and division alignment/branches misnamed rather than misorganized. Aligning with concerns that I-Dept contains functional stovepipes and integrative misalignments, some respondents opined that its branch structure is misnamed as a legacy of its inception. Observers are confused, they argued, when the names of I-Dept branches do not correspond well to their functions. Considering the actual functions of these branches and divisions absent the names, however, reveals an effective organization.

29. I-Dept undermanned for full extent of tasks and short on department/branch heads. Interview respondents currently or formerly from I-Dept and those observing it from outside often reported that I-Dept is undermanned relative to the tasks it is asked to complete.

> I-Dept staff is a small staff, underpowered, so MCIA does a lot of I-Dept stuff that in another service the I-Dept would do for itself.

During our interview period, I-Dept was also short several department or branch heads—uniformed billets that were empty.

> They don't have enough people at I-Dept. Compounding the problem, all the military billets there tend to rotate at once. They should stagger the rotations better. Worse, the senior positions rotate even faster. The good ones are gone too fast (less than a year). The civilians are good, but they need more of them. Of course, if they find someone good, they really pile work on them.

30. Many different views of MIC; confusion. During our interviews, there was considerable hesitancy to discuss the proposed MIC. This was not because respondents were not forthcoming but because few felt that they understood the MIC concept well enough to answer questions about it. Among those who did offer discussion, there was a wide range of views. As one respondent reported, "The MIC idea is not clearly articulated."

Some of the views encountered included (1) that the MIC would just be a process for tasking the intelligence battalion while in garrison—basically a streamlined request for information process; (2) that the MIC would be a manned entity and would make MEF G-2s more productive in garrison; and (3) that it is a vaguely defined reachback capability for which current infrastructure is insufficient.

31. Concerns about analysts—difficult to "make" an analyst (something intrinsic and a lot of training needed). Several respondents observed that intelligence analysis, much more than collection or production, requires both training and something intrinsic to the marine that cannot be trained. For example,

> I'm concerned about how we recruit and train analysts. We need a screening process to identify the ability to be an analyst, because most guys really don't have what it takes.

Others expressed concern about the sufficiency of training of analysts. For example,

> The schoolhouse teaches basics, but too often, marines out of the schoolhouse have no sense of analysis and no briefing and no writing skills.

Another noted,

> The 0231s we see, really, most can't be full-on analysts. They can do data mining, they can do grunt work, but there needs to be more training and development to make them into true analysts. To be a real analyst, they need a B.A. degree by the time they are a staff sergeant.

Similar concerns were raised about master analysts:

> Training MOS 0205, a master analyst, is a new challenge. How do we determine mastery? Are there criteria? How do you progress from novice to journeyman to master? No one is leaving basic school as a master. We need the right MOS roadmap.

32. Bandwidth issues at the tactical level on certain deployments and a lack of connectivity in garrison. Several respondents raised the issue of bandwidth and connectivity. We were unsurprised to learn that some isolated formations in Afghanistan had very poor connectivity and available bandwidth, and that this affected their ability to access or receive intelligence products. We were more surprised to hear that bandwidth was an issue between intelligence battalions and larger combat element formations. For example, regarding intelligence battalions,

> If I am not collocated with them, their production capability is useless to me. If not colocated, low bandwidth further forward will keep me from getting the products they make.

We were most surprised, however, to hear not about bandwidth but about connectivity problems (specifically, classified network access) for intelligence formations in garrison.

33. Concerns about I-Dept and DIRINT grade structure relative to the rest of the USMC and the broader IC. Several respondents noted that DIRINT "is a one-star in a three-star world." They reported this as a challenge both inside and outside the USMC. For example, USMC intelligence representatives in joint or interagency meetings are almost always one grade lower than their counterparts from other services—if not two. The same is true for civilians. This is also a challenge when trying to recruit and retain civilians, as comparable positions in other services' intelligence organizations have "higher pay, access, and recognition."

34. All USMC intelligence officers go from specialist to generalist. Several respondents took issue with the progression of marine intelligence officers from specialist to generalist. Under current career progressions, intelligence officers begin with a technical specialty aligned with one of the intelligence specialties. After several years of specialty service, however, all officer intelligence MOSs merge to single generalist MOSs. According to one respondent,

> Regarding retention, the generalization drives a lot of officers out. They don't want to be a generalist; they want to be a specialist and stay in the specialty. They can't, so they leave the Marine Corps.

Others reported concerns about the lack of senior officer specialists, or the lack of career paths that do not project for the possibility of full career advancement but do provide the opportunity to truly master and productively practice a specialty: "The Marine Corps doesn't want technical expertise."

35. Difficulty matching correct assignment (general support, direct support, other) with detachments; "multiple masters"/problems in general support. USMC intelligence has undertaken many initiatives to push intelligence capabilities down to the lowest tactical levels. Several respondents questioned the alignment of the variety of ways that this can be done. Intelligence personnel can be assigned to a formation, permanently, as part of their T/O. Intelligence manning in the GCE has increased substantially in this way over the past decade (see Chapter Three). Intelligence personnel at the MEF level (usually from the intelligence and radio battalions) can be temporarily assigned as part of detachments, either in direct support of a unit or in general support of a geographic area. Each of these assignments creates a different relationship between detachment marines and their "host" and "parent" formations, each with its attendant issues (especially in terms of pulling detachment marines in different directions).

Most of the concerns we heard surrounded the general support relationship, in which detachment troops are assigned to a geographic area but depend on local maneuver forces for sustainment and security.

> Our teams are not self-sustaining. They don't have the gear or logistics to do their own thing, so they rely on the infantry battalions. The infantry battalion perspective is, if you drink my water, you belong to me.

Detachments in general support are often forced to juggle instructions from their own chain of command, the desires of local maneuver commanders, and their own sustainment and security needs (which are met at the mercy of the maneuver commander). Sometimes, this is easy (all are marines and want to support the operating forces); sometimes, it is less so. According to one respondent,

> The G-3 decides attached, [direct support], or [general support]. In practice, these decisions are taken in consultation with radio battalion and intelligence battalion commanders.

Sometimes, these relationships work as planned, and sometimes they do not.

36. I-Dept lacks a single entry point and advocate for GCE, LCE, ACE. One respondent observed that there is "no customer-focused organization structure at I Dept" and that the GCE, ACE, and LCE lack an institutional advocate or clear entry portal at I-Dept. Another opined, "I-Dept needs a single entry point for GCE, ACE, LCE."

37. Need seniority and experience to conduct quality, high-level analysis and represent intelligence within and outside the USMC. Issue 14 raised the need for experience and expertise at the lowest tactical level. Other respondents raised the need for senior and expert personnel for high-level analysis and as intelligence representatives to other organizations, both inside and outside the USMC.

Several respondents reported the benefits of having senior intelligence officers as representatives:

> When we had the O-6 slots filled with O-6s, things were better. They bring experience, knowing what to make an issue of, cutting to the core of an issue.

Many respondents also noted that the USMC representative to an IC, joint, or interagency meeting is almost always lower-ranking than other services' representatives. Reportedly, such deficiencies are usually overcome by expertise, however; that is, the USMC representative may be lower-ranking, but he or she has sufficient expertise to speak with authority.

This raises a complicated question, however: "Do you want your senior and expert intelligence marines to be at the lowest tactical level, doing high-level analysis, representing Marine Corps intelligence in broader communities, or (somehow) all three?"

38. Enterprise-wide issues with knowledge management and information management. Several respondents raised concerns about knowledge management in the enterprise. One respondent went so far as to suggest that "the most pressing problem for Marine Corps intelligence is knowledge management and information management."

Another went into some detail about why it is so difficult to address:

> There is redundancy, unintended duplication, and yawning gaps. There is a tendency to look at this as strictly a technical problem, but that is not wholly the case. Problems with duplication and gaps can't be fixed by just an IT guy. Addressing these problems requires an artist to be a translator between "grunt" and "IT." That is, someone who can see the business process, identify what is wrong, then build systems to fix the processes, or fix the business process and the technical architecture at the same time.

Other respondents lamented the failure to capture many new best practices or lessons learned, including recording and learning as part of knowledge management. "If there was a better knowledge management structure, marines could possibly share their knowledge."

Another respondent offered a short, bleak summary: "We're broke with data, quite frankly."

39. Lots of informal networking (buddy network) required to get things done. Related to personality-driven command relationships (issue 21) is the reliance on the "buddy network." Respondents reported that USMC intelligence is a relatively small, tight-knit community, in which "everybody knows everybody." Because of this "family" relationship, much is accomplished not by virtue of a procedure but because of informal leveraging of the buddy network. Virtually all respondents who mentioned the buddy network discussed it as a good thing, as a way to get things done in spite of the processes. In fact, several offered anecdotes about tasks that would have been much harder, if not impossible, without relying on personal connections. According to one respondent, "Intel in the Marine Corps doesn't work by design; it works because of the people in it."

40. Limited I-Dept budget. As one respondent noted,

> One thing is critical for I-Dept: money. There is no budget for DIRINT. He has no money. MCIA is the only thing he controls with any money. I-Dept begs, borrows, and steals to pay for things, and MCIA ends up being the stuckee for that a lot, too.

The worst part is that MCIA doesn't know how much I-Dept will need, because they don't budget; they just push expenses and support requirements down (conferences, meetings, etc.).

Several other respondents noted (or lamented) I-Dept's lack of funds.

41. Grade and seniority issues relative to others in the USMC and at the interagency level. Several respondents reported that intelligence marines were often lower-ranking than those they were called on to collaborate with. This was observed both within the USMC and in the broader joint and interagency communities. Several respondents noted that the DIRINT "is a one-star in a three-star world," since other service intelligence chiefs or heads of intelligence agencies are usually three-star generals, flag officers, or equivalent. This disparity cascades across USMC intelligence representatives to IC or interagency meetings, where the USMC representative is often one or even two grades below other representatives.

Inside the USMC, intelligence grade and seniority were reported to lag behind those of colleagues in certain contexts. Several respondents explained that the operations officer (G-3) on staffs was often a grade above the intelligence officer (G-2). When they were the same grade (both majors, O-4, for example) the G-3 would be a senior major, while the G-2 would be a junior major. With this subordinate relationship, providing effective intelligence support to a commander and to operations was often reported to hinge on personalities (per issue 21).

42. Concerns about lateral entrants—insufficient screening/requirements, placement in "experienced" positions without experience. Many respondents raised concerns about lateral entrants. These concerns usually took one or both of two forms: questions about a lateral entrant's level of preparation for a certain MOS (some are underqualified or otherwise unprepared) or questions about promotions earned for time in service, allowing a senior marine who is relatively new to intelligence to serve in a billet that requires a marine with much greater intelligence-specific experience. For example,

I've seen some lat-mover captains not having the experience necessary as an 0202 to be the battalion S-2. They should be required to take and pass the MAGTF intel officer course.

43. Excessive bureaucracy in MCIA, with command and subordinate commands. A few respondents reported coordination problems internal to MCIA stemming from the organization's top-heavy structure and the fact that it has subordinate commands instead of operating as a single command. One respondent from MCIA said,

> I still struggle with MCIA as a command with three subordinate commands. So much bureaucracy. I understand how it all came about, like a typical Marine Corps command. Still, I sometimes have trouble with subordinate commanders. I wish we all worked for one guy.

Part of this issue connects to workforce and cultural issues (discussed later) regarding the need for command billets for promotion. Respondents noted that part of the reason for the subordinate commands is to provide more opportunities for intelligence officers to have commands. The problem, of course, is that "commanders want to command," even when a more streamlined organization might be more efficient.

44. In an intelligence battalion, almost no intelligence officers do intelligence work; most are consumed by administrative, management, oversight, and command tasks. This issue is phrased as a more-or-less verbatim quote from an interview respondent. Others made similar observations.

Some of this is in the nature of being a military officer, but much of it stems from obligations of command. There are related workforce and culture issues that concern the requirements of command for promotion, reflecting one of the trade-offs in this area (see also issue 46). Were intelligence officers not burdened by administrative and command responsibilities, they could do more actual intelligence work.

45. Struggle to capture or share new solutions or procedures, which leads to reinventing the wheel (includes lack of tools for capturing and sharing lessons learned). Related to knowledge management (issue 38) is an issue raised in our interviews regarding capturing innovations or other lessons learned. As one respondent said, "We do well at figuring things out, but we're bad at sharing it and storing it."

46. Command experience required for promotion. A significant cultural issue that affects the USMC intelligence officer workforce is the requirement for command experience for promotion. By design, the intelligence career field is just like other USMC career fields.

> Since 1994, the Marine Corps has built intelligence officers to do other (regular) jobs and the command positions that are needed for promotion. The Marine Corps has made intel officers look like other Marine Corps officers.

Respondents thought that this had pros and cons. On the plus side, it makes all officers equal, preventing unwarranted prejudice against or between career fields and making intelligence officers eligible for the highest promotions and career opportunities in the USMC. On the negative side, it forces intelligence officers to generalize, to meet prerequisites that have nothing to do with intelligence, and to seek command experience in order to be promoted. These requirements are directly at odds with being able to offer intelligence expertise to the broader IC:

> Marine Corps intelligence wants its people to be the equal of anyone in the IC, both based on certifications and their portfolio of work. However, by making intelligence a successful career progression, we've also shot ourselves in the foot.

47. Standards and tools to contribute to national intelligence not in place. Several respondents noted that not all USMC intelligence products comply with all the standards for reporting and recording dictated by national intelligence frameworks. Several respondents reported situations in which USMC intelligence had information that would have been useful to the broader IC, but the information either was not entered into the system (because the standards were not followed) or was not given the credibility it deserved.

Other respondents raised concerns about the ability to communicate intelligence requirements to other members of the IC:

> Many times, a requirement may be easy for someone to meet and hard for someone else to meet, but because of how requirements are (or aren't) passed, the requirement either doesn't get met or gets met at greater cost.

48. Need for command billets. Because command experience is required for promotion, even for intelligence officers, there is a greater need for command billets to satisfy workforce development needs than are actually needed for effective C2 of USMC intelligence. Respondents recognized this tension:

> Having a commander and then subcommanders can lead to tension, but if we want intel to function like a regular Marine Corps occupational field specialty, we need to have command billets.

CHAPTER SEVEN

Alternative Structures and Their Assessment

This chapter describes current aspects and structures for the four levels of USMC intelligence that we discuss in this monograph and then suggests alternative structures that address many of the issues identified in Chapter Six. These structures aim to provide better fit among USMC intelligence goals, strategy, resources and authorities, and environment. Improved fit should increase both the efficiency and effectiveness of USMC intelligence organizations. This discussion relies on terminology that was introduced in Chapter Four.

Currently, USMC intelligence organizations are not dysfunctional, but they are not without flaws and they can be improved. The current structure does serve USMC intelligence customers, but it could have better "fit" in some areas and better mitigate concerns that were raised during our interviews. As one moves from the supporting establishment to the CE and to the combat elements, the components of organizational "fit" change. At the I-Dept level, efficiency dominates as a goal, and the emphasis is on improving processes to meet institutional needs. In the combat elements, intelligence organizations should be oriented toward effective products and meeting customer needs. Strategies move from working within structured bureaucratic rules in a complex but predictable environment toward taking initiative in complex and unpredictable environments.

Our assessment of USMC and USMC intelligence strategic intent is that it emphasizes the customer and capability integration. This emphasis is one element of fit that considers organizational design. In particular, habitual interactions at all levels can improve understanding of customer needs and, on the other side, customer understanding of USMC intelligence capabilities to meet those needs. Moreover, forced integration of functional stovepipes at the producer level eases the consumer burden to integrate intelligence at the user level. Thus, if intelligence products are fused before reaching the consumer, users are saved the burden of needing to synthesize products from different intelligence sources themselves—leaving aside the fact that a trained all-source intelligence analyst should be better at such synthesis than the customer.

Organizational design must be an ongoing process as goals, strategies, resourcing, and the environment change. The *MCISR-E Roadmap* is a useful start.[1] Whatever further structural decisions are made now should be monitored and adjusted as intelligence needs evolve. In particular, current doctrine envisions a small-war environment for an extended period. If this environment changes in the future, organizational relationships might also need to change.

In this chapter, each section begins with a description of the "as-is" structure, followed by a list of concerns, a recommendation for a different structure (if applicable), and an assessment of improved fit and mitigated concerns. Our recommendation for each organization is based on a straightforward but subjective assessment of possible alternative forms. For example, we assessed how each alternative structure provides "fit" among goals, strategy, resources, and environment and addresses the concerns raised in terms of performing not well, well enough, or better. Only the structural alternative that we assessed as being best is discussed in the "should-be" section. Additional detail about our process and assessment for each alternative can be found in Appendix F.

Table 7.1 provides a summary of our assessment and recommendations.

---

[1] See U.S. Marine Corps Intelligence Department, 2010.

Table 7.1
Summary of Structural Assessment and Recommendations

Organization	As-Is structure	Concerns	Should-Be Structure
I-Dept	Functionally aligned hierarchy	Accumulation of 20 years of add-ons	Realigned functional hierarchy
MCIA	Hierarchal Functional Many customers	Excessive bureaucracy Lacks customer orientation Unclear priorities	Front-back alignment
MEF (intelligence and radio battalions)	Functional stovepipes Divisional	Focus is upward and disciplinary	Integrated matrix Habitual relationships
Combat elements	Recent shift from functional to matrix	Hampered by lack of experience	Better execution aided by change at the MEF level

Intelligence Department

As Is

The current structure of the I-Dept is a functionally aligned vertical hierarchy. The organization's broad goals are to provide input to USMC and IC policy and resource processes. There is a premium on efficient management that is typical of functional organizations. Its strategy is one of exploitation—seeking to improve functional operations in a measured way to the benefit of the USMC, its operating forces, and USMC intelligence. There is a mix of civilian and military expertise and experience, and decisionmaking is centralized, with a formal structure for coordination. However, a vertical organization needs guidance and decisions that flow quickly down, and information must flow quickly up. Recent staffing of previously vacant key leadership positions should aid in efficiently addressing such issues. The environment is complex in that a large number of factors affect the I-Dept; however, most of these factors are known.

Concerns

Concerns remain in several areas. First, there has not been a consistent, long-term, strategic focus on overall USMC intelligence goals because the various offices are more consumed with day-to-day activities. The I-Dept is more focused on inputs (e.g., money and manpower) than on customers (e.g., the operating forces). The exception is the support provided to the Commandant and USMC Headquarters staff by one operating activity. Interviewees recognized the Commandant as customer 1 for the DIRINT. Moreover, the I-Dept, while a relatively newly established organization (about ten years old), has grown in a more capricious—rather than planned—manner. As a result, names of subunits appear misaligned relative to the functions actually accomplished, and the organization is somewhat opaque to outsiders and therefore difficult to engage. Although the people assigned do have experience, at the time of our study, the military grade structure was lower than authorized, and there were a number of vacancies. In terms of authority, the DIRINT is a one-star in a headquarters world of three-star generals and is thus excluded from certain decisionmaking forums.

Should Be

The identified issues and concerns can be addressed without changing the nature of the functional structure but by aligning it in a better fashion. This includes a better grouping of the intelligence functions within one division that can then ensure coordination in addressing policy, resources, and other issues. Renaming certain subunits will improve visibility into the overall organization. A future-oriented subunit is needed, whether called a strategic initiatives group, a futures cell, or a plans shop. In addition, several of the resourcing functions might be better grouped together. Appropriate roles and reporting relationships should be established for senior civilians. One subunit with an operational orientation (the Intelligence Estimates Branch) is fundamentally different from other I-Dept subunits and could be placed elsewhere. However, because of the offices that it supports (primarily the Commandant), it should remain in the I-Dept to keep primary intelligence support to the Commandant directly under the auspices of DIRINT. Figure 7.1 presents a notional illustration of a potential realignment. In

Figure 7.1
Notional Intelligence Department Functional Hierarchy

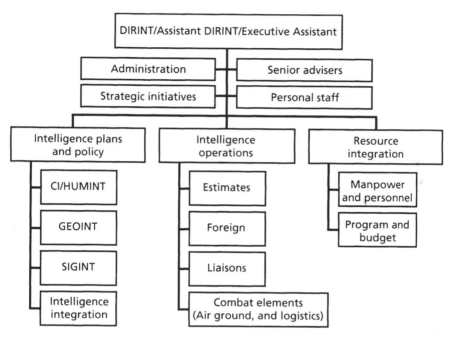

RAND *MG1108-7.1*

this structure, comparable activities are grouped by intelligence function or by broad fields of interest, such as resources. Not all activities are shown in the figure. Certain activities, such as administration and strategic initiatives, are conducted in direct response to DIRINT requests. However, this structure is but one possible realignment; in practice, the structure should reflect the DIRINT's preferences, responsibilities, and needs. In summary, the I-Dept is a functional hierarchy and should stay that way while making opportunistic improvements, such as those proposed earlier.

Improving Fit and Mitigating Specific Concerns

The recommended realignment of a functional organization enhances the as-is organization in several ways. First, it should improve the efficiency of engaging in USMC, DoD, and broader IC processes by clarifying the structure of internal organizations to improve functional coordination, make the I-Dept less opaque to outsiders, and enhance the capability for functional analysis. A significant improvement would be to provide a specific separate organizational capability for a long-term, strategic focus. These changes would be incremental from the present structure. Revolutionary change is not needed because the overall goal remains efficient outcomes with limited resources in a complex but predictable environment. Many of the concerns that were raised in our interviews focused on staffing rather than structure and, over the course of this study, the USMC addressed several of these staffing issues.

Marine Corps Intelligence Activity

As Is

MCIA currently has a mix of structures. Part of MCIA is functional and part is divisional (i.e., the military commands in the organization). MCIA is another organization that has experienced recent growth and internal changes.

MCIA has a number of goals, including producing intelligence products for a range of customers up and down the chain of command, supporting the I-Dept across DOTMLPF (doctrine, organization, training, materiel, leadership and education, personnel, and facilities) functions, being the IC's lead for cultural intelligence, coordinating IC participation, and providing a fixed site for USMC intelligence integration.[2] Its strategies appear to be twofold: efficiently produce needed intelligence products while also providing analytic leadership and innovation. MCIA is a relatively large organization with a substantial civilian and military staff. Personnel assigned are experienced, functional experts. Many of the civilian staff are former military personnel. In addition, there are clear hierarchies within the organization. MCIA exists in a complex, relatively predictable environment.

---

[2]   On of the organizational changes outlined in the *MCISR-E Roadmap* is the establishment of MCIA as the MCISR-E fixed site. This would involve MCIA's transition to a centrally managed, federated analysis and production capability, with expanded reachback capacity. It would also serve as the hub for USMC ISR data and services.

Concerns

We identified a number of concerns with regard to MCIA's structure. Mission and customer priorities are not clear. Customer service is lacking, whether through an effectively oriented web presence or 24/7 service. Products and services lack functional integration focused on customer needs. Serving multiple masters leads to complex coordination processes with resources that do not always align with priorities. The multiple customers (e.g., IC, DIRINT and I-Dept, operating forces) and MCIA's functional organization lead to frequent "reach in" by knowledgeable personnel to gain needed data, information, or assistance, often to the detriment of overall organizational functioning.

Should Be

We examined two structural alternatives for MCIA. One was a divisional form and the other was a specialized matrix form, the front-back matrix. In the divisional form, MCIA would be restructured into at least three independent units, each focused on a major set of customers. One unit would align to the IC, a second to the supporting establishment, and a third to the command and combat elements (MEF and below).[3] Each of these units would have the functional expertise needed to support its particular customer, with duplication of expertise as a result. The aim of this structure would be to more effectively support customers while recognizing a loss of the efficiency of the pure functional organization. This organizational form addresses the lack of customer focus by explicitly structuring around customers. Each customer would have the assets needed to support its anticipated needs. The head of each division would work with customers to prioritize needs and assign staff. Habitual relationships should lead to better support and even an ability to anticipate needs. It also helps to integrate intelligence functions and the intelligence cycle in a product- or customer-focused manner while recognizing that each division might do this

---

[3] Variations of this structure that could lead to more divisions than three could result from more intense customer orientation. For example, a separate division might be created for NSA relationships or separate divisions for each of the MEF (because of potentially different geographical interests) or separate divisions for GCE and ACE.

differently. On the downside, because resources would be partitioned to individual divisions, it could be difficult to flex the overall organization as customer priorities change. It would also be more difficult to maintain overall expertise across the whole organization, as training and the use of skills and resources would be the province of the divisions. Efficiency will diminish as functional resources become more decentralized. Finally, customers may compete for MCIA resources or have differing perceptions of the level of support provided. Figure 7.2 shows a notional organization chart for this alternative.

Another potential structure for MCIA is a specialized matrix form called a front-back organization. This structure is designed to accommodate both customer and product effectiveness and functional efficiency. The "front office" is a customer-focused entry and exit point for the organization. The "back office" is a functionally organized operation. Between the two is a production function unit that prioritizes requirements and assigns and coordinates fulfillment. In practice, customers seek out their point of contact and express a need.[4] The produc-

Figure 7.2
Notional MCIA Divisional Organization

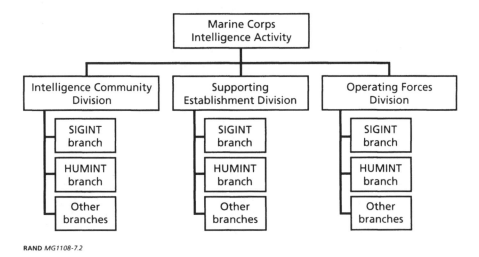

RAND *MG1108-7.2*

---

[4] There are different ways to structure the front office. One way is to have dedicated customer representatives, one person or a small group to handle requests from a given MEF or

tion function works with the functional side of the organization to task-organize to generate the needed product. The customer point of contact monitors progress and keeps the customer informed.

For example, a regimental or division S2 might interface with MCIA in several ways. An improved website could provide products that are general in nature and that are periodically updated. Specialized requests would be made by phone or electronically to a representative in the customer service division (part of front office) who consistently works with the operating forces and understands their needs. It is possible that another unit has recently made a similar request and therefore the answer would be readily available without requiring further production. If not, the request would go to the coordination and scheduling division (part of back office), whose expertise is in understanding the capabilities of the functional intelligence experts in the intelligence division. It might go to one specialized branch (a single intelligence source or specialty) or to a branch that is capable of integrating knowledge from multiple functional branches. Functional branches would be integrated as needed based on customer requirements as part of the back functionality. The scheduling division would also manage priorities, with visibility over all working requests. The customer representative, and thus the customer, would be kept aware of progress and due dates. Once complete, the product would be provided to the unit. This process would eventually allow MCIA to establish a knowledge management system to catalog and simplify intelligence production and dissemination. Figure 7.3 shows a notional organization chart for this alternative.

This structural form has the advantage of maintaining easy access and habituation with the customer but allocates expertise more efficiently and allows more functional training and development of expertise because the experts are a pooled resource. Moreover, this form

---

IC customer, for example. A second way is to have customer service groups, e.g., one specializing in IC, one for the supporting establishment (including the DIRINT and I-Dept), and one specializing in the operating forces. A third way is to have one large group, with any available member able to serve any customer. This latter approach is most efficient but requires the broadest knowledge and experience. The first approach requires deep knowledge and specialized experience.

Figure 7.3
Notional MCIA Specialized Matrix Organization

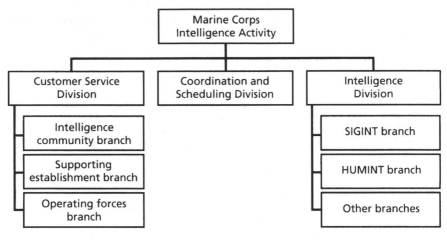

RAND *MG1108-7.3*

can better accommodate absences for training or deployment. It does require coordination between the customer representative and the production function to force functional integration of product, otherwise the default is likely to be traditional intelligence functional products.

Improving Fit and Mitigating Specific Concerns

MCIA is at a juncture regarding goals. On the one hand, efficiency of production is gained through a functional structure; on the other, a divisional alignment improves customer service effectiveness. MCIA genuinely pursues both efficiency and effectiveness. The suggested hybrid form addresses both goals.

What would this design do better than the current organization or the divisional form? First, it should more effectively produce intelligence products for customers up and down the chain of command and better support intelligence doctrine. It clarifies MCIA's role as the fixed site for intelligence across all organizational levels. It places a premium on efficient and innovative production to meet customer needs. Moreover, the structure should clarify mission and customer priorities and could lead to improvements in MCIA's web access to help it transition to a 24/7 capability as needed by certain customers. It would also allow

MCIA assets to be better used by all customers. It facilitates access by clarifying the specific means and procedures to gain it. Functional integration is focused on customer needs and tasks related to those needs. It best serves multiple masters, especially I-Dept (supporting establishment); greater transparency reveals how resources are being allocated to priority needs. Coordination becomes somewhat more complex, however, because the hierarchies are not as clear as in a divisional form.

Marine Expeditionary Forces

At the MEF level, there is a G-2 staff, but our analysis is focused on the organizational structure of the two manpower-intensive units, the intelligence battalion and the radio battalion. While there are some differences among the three intelligence and three radio battalions, the following assessment is appropriate for all. We do not explicitly address MARFOR Reserve or MARFOR Special Operations Command structure.

As Is
The battalions are organized in a divisional form, but the divisions are organized not by geography or customer but by function. These battalions are designed to collect, analyze, produce, and disseminate intelligence for use by customers up and down the chain of command. The broad strategy is to exploit existing capabilities while selectively innovating—largely as a result of technology insertion. The MEF level holds the largest set of intelligence resources, accounting, in the aggregate, for more than 50 percent of all USMC intelligence personnel.[5] In their divisional form, the battalions have independent decision-making authority, but exercising that authority frequently depends on

---

[5] The intelligence battalion is composed of a battalion headquarters, a headquarters company, a production and analysis company, a production and analysis support company, a CI/HUMINT company, and a CI/HUMINT support company. A number of sections and teams make up the company-level units. In total, there are about 75 officers and 600 enlisted personnel in the battalion. Of this number, about 550 have an intelligence MOS, either in the 02 or 26 career field, with the vast majority in 02.

personal relationships among the G-2, the major headquarters group commander, and subordinate commanders. Especially in deployed situations, they operate in a complex and unpredictable environment.

Concerns

More concerns and issues were raised about this level of intelligence organization than any other. In our interviews support of the combat elements was generally described as lacking in that it is not relevant or timely. Moreover, products are not sufficiently integrated across functions. When there are competing demands, servicing the "up" customer takes priority, irrespective of real need. The battalions are viewed as not understanding the "down" customers, such as the GCE and ACE. While teams from both the intelligence and radio battalions do provide support to the combat elements, they are limited in number, limited in the actual downward support provided, and not habitual (i.e., they lack continuity). The intelligence battalion trains as an intelligence battalion but does not deploy as a battalion, while the radio battalion is perceived as residing in its own cocoon.

Should Be

We examined two structural alternatives that used a form of matrix organization. One alternative involved a change only to the intelligence battalion, while the second involved changes to both the intelligence and radio battalions. Under both alternatives, the purpose of the significant changes was to integrate functions within the battalion by creating integrated, company-level units and to habitually associate these units with particular regimental combat units.[6] In general, we suggest restructuring the battalion into three parts: a head-

---

[6]   A model for this in the broader USMC is artillery. However, there are differences. Artillery units are one level higher than intelligence units. For example, we are proposing an intelligence company to support an RCT where, for artillery, it would be a whole battalion. Moreover, artillery has a well-developed doctrine for this support, with fire support coordinators or artillery liaison teams allocated to all levels of the supported organization. Intelligence doctrine only discusses the role of the intelligence battalion commander as the overall intelligence support coordinator for the MEF. (This has its own problems, as discussed earlier.) If a change is made to the organizational structure for intelligence at this level, doctrine needs to be extended beyond that for the intelligence support coordinator.

quarters company that would also contain manpower resources for the Intelligence Operations Center (also called a tactical fusion center) at the MEF level, a general support company for the division, and a direct support company for each of the three RCTs.[7] In the direct support company, there would be enough integrated structure to provide the teams to support the S2 of the infantry battalions in the RCT. In each direct support company, there should be at least two platoons: a CI/HUMINT platoon, focused on collection, and a production and analysis platoon. (In the second alternative, a radio battalion platoon would also be incorporated in these companies.)[8] The production and analysis platoon would have a GEOINT section and an all-source fusion section. Collection management and dissemination would occur in the company headquarters, with a small group of people trained and specializing in collection requirements management. There should be sufficient manpower so that each platoon can contain four teams. This arrangement would be the same for the general support company. Thus, there would be four teams of each type: four geospatial support teams, four production and analysis teams, and so on. For a direct support company, this would naturally mirror the supported regiment structure (one team per infantry battalion, plus one for the regiment headquarters). The S2 for the supported unit (either the RCT or the battalion) would be the intelligence support coordinator, with assistance provided by the general or direct support company commander or platoon or team leaders.[9] In practice, the USMC is familiar with such an integrated structure because it is the basic structural form of

---

[7] As discussed earlier, a more aggressive form of matrix orientation toward customer service would be to directly attach these units to the RCT for both operational and administrative control.

[8] We recognize that there are issues with this proposal in terms of classified connectivity and relationships with certain members of the IC. The trade-off is to achieve a more integrated, habitual relationship between SIGINT and the combat elements at the risk of the relationship between USMC SIGINT and the NSA.

[9] Earlier, we discussed the lack of experienced captains overall and the concentrated use of captains in levels other than the combat elements. Assigned infantry battalion S2s should be experienced captains and not junior lieutenants.

the MEU intelligence capability. Figure 7.4 shows a notional organization chart for this alternative.

There are a number of advantages to this structure. It better supports decentralized decisionmaking and, since the *Marine Corps Operating Concepts* focuses on the MEB as the key organization, it provides dedicated and habitual support for that commander.[10] Innovative and agile expertise is pushed further down to the tactical level, and intelligence functions and the intelligence cycle are integrated at a lower level (with a COIN focus) and staffed by intelligence professionals. More intelligence capability at a lower level in the COIN environment should result in intelligence moving up the chain of command that is superior to the intelligence now coming down from the higher levels.[11]

There are weaknesses in this structure, however. To be successful, a matrix structure requires expertise and experience, as well as suffi-

Figure 7.4
Notional MEF Intelligence/Radio Battalion Organization

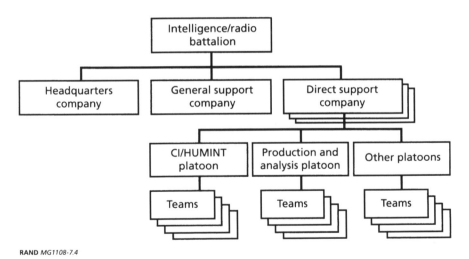

RAND *MG1108-7.4*

---

[10] See Deputy Commander for Combat Development and Integration, 2010.

[11] It is recognized that, in a large war or other kinetic environment, more intelligence resources are likely to be effective at higher levels. The commander retains the capability to move assets higher as needed, and this is arguably easier to accomplish than providing assets down if the supported unit has not habitually trained with those assets.

cient manpower to analyze and fuse intelligence at the lower levels.[12] If USMC manpower reductions occur, a strong argument needs to be made to preserve the general and direct support intelligence capability as a key component of the combat elements. Habitual support relationships might allow combat element commanders to make this argument. Moreover, the concept could serve as a model for integrating the JSF asset when it becomes available. Last, in any matrix structure, the multiple masters problem will remain. In this case, the intelligence or radio battalion commander has administrative control of the general and direct support, while the RCT has operational control. This could be ameliorated by having RCT or infantry battalion S2s serve as doctrinal intelligence support coordinators, somewhat removing the intelligence or radio battalion commander from the immediate tactical situation.

Matrix structures can be unwieldy and inefficient, but they are how the USMC currently task-organizes. This could be the organizational fit that explicitly aligns intelligence structure to the way in which the USMC does business, especially given the 2010 *Marine Corps Operating Concepts* discussed earlier.

Improving Fit and Mitigating Specific Concerns

The advantages discussed here mitigate many of the important concerns that were raised during our interviews. This structure better supports the combat elements through direct support and habitual relationships. Moreover, it focuses priorities on the combat elements while still satisfying higher-headquarters needs. In garrison, continuing relationships with a specific unit in one of the combat elements should lead to better training and more useful intelligence products, as well as a better understanding of intelligence capabilities by commanders at all levels.

---

[12] The discussion in Chapter Three shows how the intelligence workforce is currently skewed toward junior grades relative to other USMC career fields and explains why we believe that this situation will correct itself over time.

Combat Elements

As Is

Our review of the GCE indicates that a structural shift has already occurred and the result is better than the previous organization. Over the past few years, the intelligence structure of the GCE has shifted from functional to matrix—from a battalion-level functional S2 intelligence structure to a company-level intelligence cell in which intelligence personnel from the S2 section are matrixed with infantry marines at the company level. This has allowed intelligence marines to be allocated to the lowest possible level (considered essential in COIN operations) while retaining the flexibility to reallocate personnel based on variations in demand.

Concerns

The liability that exists for a matrix structure in the present environment is that intelligence personnel assigned to the battalion level need to be experienced and expert in their craft, and that is not always the case. Greater availability of experienced personnel for the CE in the direct support model could help mitigate this concern. Moreover, as recently recruited and trained intelligence personnel gain expertise, assignments at this level will cultivate more experienced personnel.

Should Be

Our assessment is that, especially for the current COIN environment, this is an appropriate blend of effectiveness in employment that maintains S2 involvement in efficiently training personnel and overseeing their use.

Summary

The structural alternatives recommended in this chapter should improve the efficiency and effectiveness of support to leaders and operational forces by providing the information that they need. The alternative structures improve effectiveness by leading marines to collect and deliver information that is more relevant for consumers because they foster a better understanding of what consumers want and need. The alternative structures also improve efficiency by preserving functional competencies while prioritizing customer needs in a manner that aligns with strategic intent.

CHAPTER EIGHT

Addressing Remaining Marine Corps Intelligence Issues

The organizational changes proposed in Chapter Seven sought to improve the fit of various elements of USMC intelligence. This chapter aims to connect the analyses in the previous two chapters and offer resolution or prospects for solutions to as many of the issues identified in Chapter Six as possible. We begin by considering which of the issues identified in Chapter Six would be resolved by the changes proposed in Chapter Seven. We then turn to solutions and suggestions for improvement offered by interview respondents and discuss their relevance to the other issues.

Issues Affected by Structural Change

Nineteen issues would be resolved, or at least partially addressed, by one or more of the organizational changes proposed in Chapter Seven:

- General propensity to respond to the commanding general's curiosity rather than tactical force needs
- MCIA lacks a 24-hour watch cycle
- Regurgitation, not analysis, from certain layers in the organization
- Stovepiped intelligence specialty "tribes"—lack of integration or desire to build single intelligence-only targeting profiles
- Intelligence officers who do not understand tactical context and thus provide limited value

- In current operations, the majority of manpower is in the intelligence battalion, but the majority of information is in the infantry battalion
- Issues with MCIA websites
- Seniority and experience needed at the lowest tactical level
- Flexibility needed to push personnel out to the tactical level and pull them back if needs change or if they are being misused
- Integration, coordination, and focus lacking—I-Dept very stovepiped
- MCIA serves multiple masters: USMC Headquarters, operating forces, DIRINT, IC, MCCDC; struggles to say no; lacks boundaries in terms of mission and lacks clear mission statement
- In garrison, the intelligence battalion does minimal intelligence work
- I-Dept consumed with treading water; struggles to look past immediate firefighting to long-term vision
- Impositions and informal tasking from I-Dept to MCIA
- The intelligence battalion trains as a battalion but does not deploy as a battalion
- I-Dept legacy branch and division alignment/branches misnamed rather than misorganized
- I-Dept undermanned for full extent of tasks and short on department/branch heads
- Need seniority and experience to conduct quality, high-level analysis and represent intelligence within and outside the USMC
- Lots of informal networking (buddy network) to get things done.

The remainder of this section assesses the impact of the proposed changes on each of the issues listed above.

General propensity to respond to commanding general's curiosity rather than tactical force needs. The changes proposed at the MEF level, where intelligence and radio battalions become integrated matrix organizations with habituated relationships with RCTs, should help the elements that are pushed out in service to the RCTs better focus on the RCTs. To the extent that this issue remains unresolved, executive intervention by commanding generals—making the tactical forces the priority and placing real constraints on the resources to be consumed—is the only way to mitigate it further.

MCIA lacks a 24-hour watch cycle. Although this not an explicit part of either possible MCIA reconfiguration, it is likely to improve under both approaches. With customer-oriented divisions, divisions should identify which customers require such support and establish that support. Under a front-back hybrid business model, MCIA could add a 24/7 watch as part of its front office.

Regurgitation, not analysis, from certain layers in the organization. Moving to a matrix structure at the MEF level should help mitigate this issue. Habituated relationships should both make clear and help prevent the continuation of layers that are not useful. Elsewhere in the enterprise, the adoption of a more customer-focused structure at MCIA (either divisional or front-back) should mitigate this as well by better aligning production with customer needs.

Stovepiped intelligence specialty "tribes"—lack of integration or desire to build single intelligence-only targeting profiles. Moving away from intelligence specialty tribe divisional structures and toward matrix structures (as proposed for both the MEF formations and MCIA) should help break this down trend and thus mitigate this issue.

Intelligence officers who do not understand tactical context and thus provide limited value. Establishing habitual relationships between the RCTs and the reorganized elements of the intel battalions should help. Such relationships are beneficial for exposure on both sides. Further, several interview respondents suggested requiring certain experience or credentials for certain billets. Doing so would reduce the chances that an intelligence officer could end up in a position for which he or she is unprepared.

In current operations, the majority of manpower is in the intel battalion, but the majority of information is in the infantry battalion. The matrix structure proposed for the MEF elements, in which intelligence and radio battalions are reorganized and form habitual relationships with RCTs, should significantly mitigate concerns in this area.

Issues with MCIA websites. Although this is not really an organizational issue, per se, the additional customer focus that would stem from either alternative proposed for MCIA should address this concern.

Seniority and experience needed at the lowest tactical level. The proposed matrix structure for the MEF elements and the associated habitual relationships with RCTs should help ensure the availability of experienced intelligence personnel to lower levels.

Flexibility needed to push personnel out to the tactical level and pull them back if needs change or if they are being misused. While the proposed matrix structure makes intelligence and radio battalion marines available at the lowest levels and establishes habitual relationships with the supported combat elements, control of those personnel and resources remains centralized, and they can be task-organized to support other formations or central functions should operations demand it.

Lack of integration, coordination, and focus—I-Dept very stovepiped. The proposed functional realignment of I-Dept, coupled with executive attention, can resolve or at least substantially mitigate this issue.

MCIA serves multiple masters: USMC Headquarters, operating forces, DIRINT, IC, MCCDC; struggles to say no; lacks boundaries in terms of mission and lacks clear mission statement. Reorganizing MCIA into customer-focused divisions or a front-back matrix organization should help mitigate this issue. Either structure should enforce prioritization and make clear when a "no" answer might be appropriate. Furthermore, either structure might better encourage customers to set their own priorities and would provide better tools or relationships to help them do so.

In garrison, the intelligence battalion does minimal intelligence work. Through their habitual relationships with the RCTs, the assigned companies of the reformed intelligence battalions can train with their respective RCTs to support future operations, contingencies, and deployments. While this structural change can help mitigate the current situation, the related issue of connectivity in garrison must also be resolved.

I-Dept consumed with treading water; struggles to look past immediate firefighting to long-term vision. The proposed functional realignment should mitigate this issue to some extent. As noted, I-Dept planning should be allowed more breathing room to address long-term issues.

Impositions and informal tasking from I-Dept to MCIA. This issue can be resolved by formalizing and insisting on single entry points for customers. Such a change is built in to both proposed organizational structures for MCIA.

The intelligence battalion trains as a battalion but does not deploy as a battalion. The reformed structure of the intelligence battalions and the habitual relationships of their elements with the RCTs should significantly change how the intelligence battalion functions in garrison, better aligning it with how the battalion functions while deployed.

I-Dept legacy branch and division alignment/branches misnamed rather than misorganized. This issue wholly disappears if I-Dept's functional structure is realigned as proposed. If, however, the as-is structure is retained, there would be some small benefit in renaming the divisions and branches.

I-Dept undermanned for full extent of tasks and short on department/branch heads. The proposed functional realignment should relieve some of the pressure on the I-Dept workforce. However, there is still a need to man I-Dept to T/O with the appropriate corresponding grade structure.

Need seniority and experience to conduct quality, high-level analysis and represent intelligence within and outside the USMC. Manning all the senior billets at I-Dept will help. The proposed changes at MCIA should help, too.

Lots of informal networking (buddy network) to get things done. The increased formalization and habituation throughout the enterprise that is implied by the various proposed structural changes should help somewhat. It is unclear what else would need to be done to address this issue.

Issues Not Addressed by Structural Change

The remaining 29 issues are not specifically addressed by the proposed structural changes. They are listed here, followed by our interviewees' suggestions for addressing them:

- Intelligence personnel doing collateral duties instead of intelligence
- Vicious cycle in aviation: intelligence not well prepared to support aviators; aviators view intelligence as irrelevant
- Misuse of reconnaissance or intelligence assets by controlling maneuver forces
- Layers of bureaucracy/layers of intelligence, no clear guidance on what is to be done at different levels
- PIRs are not updated often enough
- Collection requirements management gets neglected
- Authority conflated between intelligence battalion commanding officer and MEF G-2; subordinates can be pulled in different directions, and priority follows the fitness report
- Some tactical commanders do not understand intelligence
- Personality-driven command relationships
- DIRINT has limited direct authority
- Need to man for increased collection assets in ACE (i.e., JSF, UAS)
- SIGINT personnel useless without their tools and database access
- Many different views of MIC; confusion
- Concerns about analysts—difficult to "make" an analyst (something intrinsic and a lot of training needed)
- Bandwidth issues at the tactical level on certain deployments and a lack of connectivity in garrison
- Concerns about I-Dept and DIRINT grade structure relative to the rest of the USMC and the broader IC
- All USMC intelligence officers go from specialist to generalist
- Difficulty matching correct assignment (general support, direct support, other) with detachments; "multiple masters"/problems with general support

- I-Dept lacks a single entry point and advocate for GCE, LCE, and ACE
- Enterprise-wide issues with knowledge management and information management
- Limited I-Dept budget
- Grade and seniority issues relative to others in the USMC and at the interagency level (e.g., 1-star in a 3-star world; X2 always junior to X3)
- Concerns about lateral entrants—insufficient screening/requirements, placement in "experienced" positions without experience
- Excessive bureaucracy in MCIA, with command and subordinate commands
- In an intelligence battalion, almost no intelligence officers do intelligence work; most are consumed by administrative, management, oversight, and command tasks
- Struggle to capture or share new solutions or procedures, which leads to reinventing the wheel (includes lack of tools for capturing and sharing lessons learned)
- Command experience required for promotion
- Standards and tools to contribute to national intelligence not in place
- Need for command billets.

Interview respondents often offered possible solutions to the issues we discussed; where appropriate, we include those suggestions here. Note that these suggestions have not been subjected to any form of rigorous analysis or validation, save having face validity and making sufficient sense to the project team to be included. If an issue would not be resolved or mitigated by a proposed organizational change, and either no suggestions or no reasonable suggestions for its mitigation were offered in the interviews, we simply report it as not addressed.

Intelligence personnel doing collateral duties instead of intelligence. This would not be resolved by any of the proposed structural changes (though the envisioned habitual relationships between elements of the intelligence battalions and RCTs should create positive incentives to use intelligence personnel primarily for intelligence). This is fundamentally a leadership and training issue.

Vicious cycle in aviation: intelligence not well prepared to support aviators; aviators view intelligence as irrelevant. Not addressed by the proposed changes.

Misuse of reconnaissance or intelligence assets by controlling maneuver forces. Not addressed by the proposed changes.

Layers of bureaucracy/layers of intelligence, no clear guidance on what is to be done at different levels. Several respondents suggested this would be a prime target in updated intelligence doctrine, which could lay out clear guidelines for each layer along with formal rules for the allocation of intelligence manpower and assignment of tasks. Whether in doctrine or some other form of guidance, if the MEF elements move to the proposed matrix structure, there will need to be clear, general guidance on different ways to parcel out personnel and what roles they might perform.

PIRs are not updated often enough. Not addressed by the proposed changes.

Collection requirements management gets neglected. Not addressed by the proposed changes. Perhaps a small group of people trained in the discipline of collection requirements management could help meet this need.

Authority conflated between intelligence battalion commanding officer and MEF G-2; subordinates can be pulled in different directions, and priority follows the fitness report. Many respondents viewed this as an important problem, and many possible solutions were offered:

- Give the MEF G-2 administrative and operational control of the intelligence battalion when deployed.
- Make the MEF deputy G-2 the intelligence battalion commander.
- Make the intelligence battalion commander a force provider and administrator only (i.e., does not deploy).

While each of these approaches has pros, cons, or impracticalities that would be immediately evident to a USMC intelligence audience, one thing is clear from our discussions: When a marine feels that he or she is called on to serve multiple masters, he or she will try to make everyone happy, but when push comes to shove, priority will go to the officer who writes the fitness report. The solution, then, is to ensure that the officer who is expected to be in charge in these situations is the one who is writing the fitness reports.

Some tactical commanders do not understand intelligence. Not addressed by the proposed changes. The habitual relationships inherent in the proposed changes to the MEF elements should help somewhat, but there is a strong professional military education element to this issue.

Personality-driven command relationships. Not addressed by the proposed changes.

DIRINT has limited direct authority. Respondents reported that this is unlikely to change. To realize change and offer leadership through indirect authority, DIRINT must continue to rely on persuasion, cooperation, and influence.

Need to man for increased collection assets in ACE (i.e., JSF, UAS). This issue is not addressed because, currently, there are too many uncertainties to do so with confidence. Nonetheless, we recognize that this will be a high-priority concern for the USMC intelligence enterprise as manning needs solidify and become clarified. Perhaps a small group of people trained in the discipline of collection requirements management would meet this need.

SIGINT personnel useless without their tools and database access. Not addressed by the proposed changes.

Many different views of MIC; confusion. This issue can be addressed by clarifying the intent and nature of the MIC and promoting a single vision.

Concerns about analysts—difficult to "make" an analyst, (something intrinsic and a lot of training needed). Interview respondents observed that there are two components to this issue. First, when it comes to the possibility of becoming an analyst, some marines have the potential and some just don't. Second, even among those with the inherent potential, significant training and investment is necessary to develop it. The solutions proposed followed these two threads. Several respondents suggested employing screening and selection criteria for analyst MOSs to avoid wasting training on those who lack the potential to become good analysts. Others suggested the addition of mid-career training courses for analyst MOSs.

Bandwidth issues at the tactical level on certain deployments and a lack of connectivity in garrison. Not addressed by the proposed changes.

Concerns about I-Dept and DIRINT grade structure relative to the rest of the USMC and the broader IC. Several respondents suggested increasing the USMC intelligence grade structure, including making DIRINT a 2- or 3-star position. One possible counter-concern was raised that had some credibility: It is reasonably likely that if DIRINT were a 3-star billet, it would not be filled by an intelligence professional. Several respondents noted that DIRINT "punches above his weight" because he is a knowledgeable and experienced intelligence professional in his own right.

All USMC intelligence officers go from specialist to generalist. Respondents with whom we discussed this issue offered several possible alternatives, all with various pros and cons. They included the following:

- Delay the intelligence officer merge to generalist, and let specialists stay in their specialties longer.
- Do not force intelligence officers to merge to generalist; let specialists stay in their specialties with the understanding that it will constrain their promotion possibilities.
- Bring back limited-duty officers (or more warrant officers) for the intelligence officer specialties.
- Start intelligence officers as generalists (0202) from the outset; bring back limited-duty officers and warrant officers for specialist needs.

If this issue is genuinely viewed as a concern, it is an area ripe for further research.

Difficulty matching correct assignment (general support, direct support, other) with detachments; "multiple masters"/problems in general support. Not addressed by the proposed changes.

I-Dept lacks a single entry point and advocate for GCE, LCE, and ACE. Unless I-Dept shifts to a divisional structure organized by customer, this would not be addressed.

Enterprise-wide issues with knowledge management and information management. Not addressed by the proposed changes.

Limited I-Dept budget. Not addressed by the proposed changes.

Grade and seniority issues relative to others in the USMC and at the interagency level (e.g., 1-star in a 3-star world; X2 always junior to X3). Not addressed by the proposed changes except as briefly discussed with regard to I-Dept and DIRINT, specifically.

Concerns about lateral entrants—insufficient screening/requirements, placement in "experienced" positions without experience. Several respondents proposed rather straightforward solutions to this challenge. These included establishing requirements for intelligence MOS lateral entrants and ensuring minimum thresholds of MOS experience for certain assignments.

Excessive bureaucracy in MCIA, with command and subordinate commands. Several respondents offered suggestions to address this issue. The suggestions followed from the perceived reason for the excessive bureaucracy in the organization: Flatten (reduce vertical differentiation in) MCIA such that it becomes a single command with no subcommands (other than MCSB). Such an adjustment would have pros and cons. While this change would indeed address the bureaucracy issue, having a single commander for both the front and back of the proposed matrix structure would also mitigate a possible weakness in that model. However, such a change would reduce the overall number of intelligence command billets available, a trade-off in tension with some of the other issues raised.

In an intelligence battalion, almost no intelligence officers do intelligence work; most are consumed by administrative, management, oversight, and command tasks. This would not be addressed by the proposed changes, but this issue is part of the hidden cost of having intelligence command billets and structuring intelligence like any other USMC career field.

Struggle to capture or share new solutions or procedures, which leads to reinventing the wheel (includes lack of tools for capturing and sharing lessons learned). Not addressed by the proposed changes.

Command experience required for promotion. Not addressed by the proposed changes. However, this is an issue that could be addressed through the officer promotion system. Each service promotes officers slightly differently, with the Air Force and USMC most similar in that almost all officers compete against each other in a single, large competitive category. In the Army, MI (a branch) and Strategic Intelligence (a functional area) are part of the Operations Support competitive category. Also in that competitive category are signals, foreign area officers, space, academy professors, operations research and systems analysts, force management, nukes, strategic operations, strategic plans and policy. In the Navy, intelligence is its own special-duty officer competitive category. So, service promotion practices for intelligence careers span the spectrum, from intelligence as its own competitive category to personnel competing with everyone but the professionals. It is not likely that the USMC will change this practice. However, each service, including the USMC, uses promotion board precepts to articulate officer shortages or the need for special considerations, such as lack of command billets. USMC intelligence could pursue such considerations.

Standards and tools to contribute to national intelligence not in place. Not addressed by the proposed changes.

Need for command billets. Not addressed by the proposed changes.

CHAPTER NINE

Conclusions and Recommendations

USMC intelligence has come a long way since the 1994 Intelligence Plan, effecting many important changes and improvements suggested in that reform effort and making further adjustments in response to the operational context encountered in Iraq and Afghanistan. The USMC intelligence enterprise is highly effective and has realized many significant successes. However, issues, challenges, and room for improvement remain.

Findings

The Intelligence Department Reflects an Accumulation of 20 Years of Organizational Change

There has been an inconsistent long-term strategic focus on overall intelligence goals; the various I-Dept offices are more consumed with day-to-day activities. The I-Dept, by virtue of its headquarters placement, focuses more on inputs (e.g., money, manpower) than on customers (e.g., the operating forces). Moreover, the I-Dept has grown rapidly and reactively rather than in a planned manner. As a result, names of subunits do not reflect the functions actually performed, and the organization is somewhat opaque to outsiders and thus difficult to engage.

Our recommendations address these issues by encouraging a functionally aligned organization to achieve efficiency while more logically grouping the substructures by function or by broad interest.

The Marine Corps Intelligence Activity Lacks Customer Orientation and Has Unclear Priorities

Customer service is lacking, and MCIA has neither an effectively oriented web presence nor 24/7 service. Products and services lack functional integration with a focus on customer needs. Serving multiple masters complicates coordination processes, and resources do not always align with priorities. The multiple customers (e.g., IC, DIRINT and I-Dept, operating forces) and functional organization lead to frequent "reach-in" by knowledgeable personnel to gain needed data, information, or assistance—to the detriment of overall organizational functioning.

Our recommendation for a specialized matrix form of organization gains the efficiency of functional groupings while achieving the effectiveness of improved customer support.

The Focus of the Marine Expeditionary Force Is "Up" and Disciplinary

In our interviews, support of the combat elements was generally described as lacking in that it is not relevant or timely. Moreover, products are not sufficiently integrated across functions. When there are competing demands, servicing the "up" customer takes priority, irrespective of real need. The intelligence battalion trains as an intelligence battalion but does not deploy as a battalion, while the radio battalion is perceived as residing in its own cocoon.

Our recommendation is to restructure these battalions to better serve the ground, air, and logistics combat elements by moving them from a macro focus (which may be optimal for large wars) to a micro focus, which is optimal for small wars. If the environment were to change, there would be sufficient flexibility to return to the macro focus. It is easier to pull distributed forces back "up" to meet a contingency than the other way around.

The Combat Elements Have Recently Shifted from a Functional to a Matrix Structure but Are Hampered by a Lack of Experience

Over the past few years, the intelligence structure in the GCE has shifted from functional to matrix, from a battalion-level functional

S2 intelligence structure to a company-level intelligence cell, in which intelligence personnel from the battalion S2 section are matrixed with infantry marines at the company level. The liability of a matrix structure in the present environment is that the intelligence personnel assigned to battalion level need to be experienced and expert in their craft, and that is not always the case.

Changing organizational structure alone will not resolve this issue. However, as USMC intelligence personnel become more experienced and more highly graded over time, the expectation is that greater expertise will reside at the combat element level. Moreover, the habitual association of a direct support intelligence company with an RCT should also improve the level of experience and expertise in the combat elements.

Other Issues Relate to Mission, Workforce, Leadership, Culture, and Technology

Some of these issues might be construed as "organizational" in a broader sense; others, not. We discuss them throughout this monograph because they have the potential to affect USMC intelligence strategic objectives and thus may require attention or resolution through organizational changes or other approaches. Organizational change could remedy these issues, or it could be counterproductive and hamper the effectiveness of the organizational changes analyzed in Chapter Seven. Our recommendations are based mainly on the ability of the proposed alternatives to achieve organizational fit—that is, the best structural form for the environmental and strategic demands of these organizations.

Recommendations

This monograph examined the organizational structure and a collection of issues hindering the effectiveness and efficiency of the USMC intelligence enterprise on four levels: I-Dept, MCIA, the MEF (specifically, the intelligence and radio battalion), and the combat elements.

In Chapter Seven, we recommended organizational changes to three of those levels, summarized here.

The Intelligence Department Is a Functional Hierarchy and Should Stay That Way While Making Opportunistic Improvements

The issues and concerns that we identified in I-Dept can be addressed without changing the nature of the department's functional structure, but rather by realigning it. Specifically, several of the resourcing functions could be grouped together. Appropriate roles and reporting relationships should be established for senior civilians. One subunit with an operational orientation (the Intelligence Estimates Branch) could be placed elsewhere because it is functionally different from all other subunits. However, because it supports high-level offices (primarily, the Commandant of the Marine Corps), it is best kept in the I-Dept.

The Marine Corps Intelligence Activity Should Reorganize into a Specialized Matrix Known as a Front-Back Organization

For MCIA, we recommend a structural alternative that is a specialized matrix form called a front-back organization. This structure is designed to accommodate both customer and product effectiveness and functional efficiency. It can also better accommodate absences for training or deployment. Furthermore, it has the advantage of maintaining easy access and habituation with customers but allocates expertise more efficiently, and it allows more functional training and development of expertise because experts are a pooled resource. The ability to manage and monitor customer needs and demands, and to efficiently allocate expertise and resources to meet those demands, is particularly important to MCIA, with its host of varied customers.

The Marine Expeditionary Force Could Be More Effective if Organized into Integrated Matrix Habitual Relationships

A significant change at the MEF level would be to integrate functions in the battalion by creating discipline-integrated, company-level units and to associate these units habitually in both general and direct support relationships with particular regimental combat units.[1] In practice, the USMC is familiar with such an integrated structure because it is used elsewhere and is the basic structural form for MEU intelligence capabilities. This structure better supports decentralized decisionmaking and, because the USMC operating concept focuses on the MEB as the key organization, it provides dedicated and habitual support for that commander.

These recommendations should improve the extent to which USMC intelligence organizations fit with and correspond to the imperatives of their environmental context and achieve the service's strategic intent.

---

[1] A model for such habituation between formations in the USMC is artillery. However, there are differences in the traditional relationship between artillery and the regimental combat units and what we are proposing for intelligence. Artillery units are one level above intelligence units; we are proposing an intelligence company to support an RCT where, for artillery, it would be a whole battalion. Moreover, artillery has a well-developed doctrine for this support, with fire support coordinators or artillery liaison teams allocated to all levels of the supported organization. Intelligence doctrine only discusses the role of the intelligence battalion commander as the overall intelligence support coordinator for the MEF. (This has its own problems.) If the organizational structure is changed at this level, intelligence doctrine needs to be extended beyond that for the intelligence support coordinator.

Organizational Design Literature Considered

This appendix presents the sources considered in our review of the literature on organizational design. A complete list of sources cited in the text can be found at the end of this monograph.

Agarwal, Rajshree, and Constance E. Helfat, "Strategic Renewal of Organizations," *Organization Science*, Vol. 20, No. 2, March–April 2009, pp. 281–293.

Aggarwal, Gunjan, "Driving the Strategic Agenda in Research Organizations," *OD Practitioner*, Vol. 39, No. 3, 2007, pp. 46–52.

Aguilera, Ruth V., Igor Filatotchev, Howard Gospel, and Gregory Jackson, "An Organizational Approach to Comparative Corporate Governance: Costs, Contingencies, and Complementarities," *Organization Science*, Vol. 19, No. 3, May–June 2008, pp. 475–492.

Albrecht, Karl, *The Power of Minds at Work: Organizational Intelligence in Action*, New York: AMACOM, 2003.

Augier, Mie, and David J. Teece, "Dynamic Capabilities and the Role of Managers in Business Strategy and Economic Performance," *Organization Science*, Vol. 20, No. 2, March–April 2009, pp. 410–421.

Azriel, Jay, "Organizational Structure and Controls," lecture notes from Strategic Management course (No. BMGT 481), University at Albany, State University of New York, Fall 1999.

Barney, Jay B., and Delwyn N. Clark, *Resource-Based Theory: Creating and Sustaining Competitive Advantage*, New York: Oxford University Press, 2007.

Brunner, David James, Bradley R. Staats, Michael L. Tushman, and David M. Upton, *Wellsprings of Creation: How Perturbation Sustains Exploration in Mature Organizations*, Cambridge, Mass.: Harvard Business School, Working Paper No. 09-011, June 15, 2009. As of February 12, 2009:
http://www.hbs.edu/research/pdf/09-011.pdf

Burton, Richard M., Gerardine DeSanctis, and Børge Obel, *Organizational Design: A Step-by-Step Approach*, Cambridge, UK: Cambridge University Press, 2006.

Caldart, Adrián A., and Joan Enric Ricart, *A Formal Evaluation of the Performance of Different Corporate Styles in Stable and Turbulent Environments*, Barcelona, Spain: IESE Business School, University of Navarra, Working Paper No. 621, March 2006.

Christensen, Clayton M., and Michael E. Raynor, *The Innovator's Solution: Creating and Sustaining Successful Growth*, Boston, Mass.: Harvard Business School Press, 2003.

Christensen, Michael, and Thorbjørn Knudsen, "Design of Decision-Making Organizations," *Management Science*, Vol. 56, No. 1, January 2010, pp. 71–89.

Davenport, Thomas H., Jeanne G. Harris, and Robert Morrison, *Analytics at Work: Smarter Decisions, Better Results*, Boston, Mass.: Harvard Business School Press, 2010.

DiMaggio, Paul J., and Walter W. Powell, "The Iron Cage Revisited: Institutional Isomorphism and Collective Rationality in Organizational Fields," *American Sociological Review*, Vol. 48, No. 2, April 1982, pp. 147–160.

Donaldson, Lex, "Strategy and Structural Adjustment to Regain Fit and Performance: In Defense of Contingency Theory," *Journal of Management Studies*, Vol. 24, No. 1, January 1987, pp. 1–24.

Fukuyama, Francis, and Abram N. Shulsky, *The "Virtual Corporation" and Army Organization*, Santa Monica, Calif.: RAND Corporation, MR-863-A, 1997. As of March 23, 2011:
http://www.rand.org/pubs/monograph_reports/MR863.html

Galbraith, Jay R., "Organization Design: An Information Processing View," *Interfaces*, Vol. 4, No. 3, 1974, pp. 28–36.

———, *Designing Organizations: An Executive Guide to Strategy, Structure and Process*, San Francisco, Calif.: Jossey-Bass, 2002.

———, *Designing Matrix Organizations That Actually Work*, San Francisco, Calif.: Jossey-Bass, 2009.

Gortner, Harold, Julianne Mahler, and Jeanne Bell Nicholson, *Organization Theory: A Public Perspective*, 2nd ed., Fort Worth, Tex.: Harcourt Brace College Publishers, 1997.

Gulati, Ranjay, and Phanish Puranam, "Renewal Through Reorganization: The Value of Inconsistencies Between Formal and Informal Organization," *Organization Science*, Vol. 20, No. 2, March–April 2009, pp. 422–440.

Hambrick, Donald C., "New Directions in Corporate Governance Research," *Organization Science*, Vol. 19, No. 3, May–June 2008, pp. 381–385.

Harris, Elwyn D., Michael V. Hynes, Harry J. Thie, Robert M. Emmerichs, Malcolm MacKinnon, Denis Rushworth, Brian Nichiporuk, John E. Peters, Maurice Eisenstein, Jennifer Sloan McCombs, Charles Lindenblatt, and Charles Cannon, *Transitioning NAVSEA to the Future: Strategy, Business, and Organization*, Santa Monica, Calif.: RAND Corporation, MR-1303-NAVY, 2002. As of March 23, 2011:
http://www.rand.org/pubs/monograph_reports/MR1303.html

Harris, Jeanne G., Elizabeth Craig, and Henry Egan, "How to Organize Your Analytical Talent," *Analytics*, January–February 2010, pp. 15–21.

Harrison, Michael I., *Diagnosing Organizations: Methods, Models, and Processes*, Thousand Oaks, Calif.: Sage Publications, 2005.

Kaplan, Robert S., and David P. Norton, "Mastering the Management System," *Harvard Business Review*, January 2008.

Ketokivi, Mikko, Roger Schroeder, and Virpi Turkulainem, *Organizational Differentiation and Integration: A New Look at an Old Theory*, Working Paper No. 2006/2, Espoo, Finland: Department of Industrial Engineering and Management, 2006.

Kiechel, Walter III, "Seven Chapters of Strategic Wisdom," *Strategy+Business*, No. 58, February 23, 2010.

King, Brayden G., Teppo Felin, and David A. Whetten, "Finding the Organization in Organizational Theory: A Meta-Theory of the Organization as a Social Actor," *Organization Science*, Vol. 21, No. 1, January–February 2010, pp. 290–305.

Laseter, Tim, "An Essential Step for Corporate Strategy," *Strategy+Business*, Winter 2009, No. 57, November 24, 2009.

Lawler, Edward E., *From the Ground Up: Six Principles for Building the New Logic Corporation*, San Francisco, Calif.: Jossey-Bass, 1996.

———, "Rethinking Organization Size," *Organizational Dynamics*, Fall 1997, pp. 24–35.

Miles, Raymond E., and Charles C. Snow, *Organizational Strategy, Structure, and Process*, Stanford, Calif.: Stanford University Press, 2003.

Nadler, David, and Michael Tushman, *Competing by Design: The Power of Organizational Architecture*, New York: Oxford University Press, 1997.

O'Leary, Michael Boyer, and Mark Mortensen, "Go (Con)figure: Subgroups, Imbalance, and Isolates in Geographically Dispersed Teams," *Organization Science*, Vol. 21, No. 1, January–February 2010, pp. 115–131.

Porter, Michael E., "What Is Strategy?" *Harvard Business Review*, November–December 1996, pp. 61–78.

Raisch, Sebastain, Julian Birkinshaw, Gilber Probst, and Michael L. Tushman, "Organizational Ambidexterity: Balancing Exploitation and Exploration for Sustained Performance," *Organization Science*, Vol. 20, No. 4, July–August 2009, pp. 685–695.

Tushman, Michael L., and Charles A. O'Reilly. "Ambidextrous Organizations: Managing Evolutionary and Revolutionary Change," *California Management Review*, Vol. 38, No. 4, Summer 1996, pp. 8–30.

Wallace, Joseph, James Hunt, and Christopher Richards, "The Relationship Between Organisational Culture, Organisational Climate and Managerial Values," *International Journal of Public Sector Management*, Vol. 12, No. 7, 1999, pp. 548–564.

Wasserman, Noam, "Revisiting the Strategy, Structure, and Performance Paradigm: The Case of Venture Capital," *Organization Science*, Vol. 19, No. 2, March–April 2008, pp. 241–259.

Williams, Charles, "Comparing Evolutionary and Contingency Theory Approaches to Organizational Structure," in Richard M. Burton, Bo H. Eriksen, Dorthe Døjbak Hakonsson, Thorbjørn Knudsen, and Charles C. Snow, eds., *Designing Organizations: 21st Century Approaches*, Boston: Springer, 2008.

Wilson, James Q., *Bureaucracy: What Government Agencies Do and Why They Do It*, New York: Basic Books, 1989.

Woolley, Anita Williams, "Means vs. Ends: Implication of Process and Outcome Focus for Team Adaptation and Performance," *Organization Science*, Vol. 20, No. 3, May–June 2009, pp. 500–515.

Army Intelligence Organization

With some important differences, the U.S. Army and the USMC overlap in significant ways in terms of tasks and organization. It is hard to get too far into a discussion of the way the USMC organizes for intelligence and possible alternative structures without considering how the Army organizes for intelligence. This appendix summarizes the Army's current intelligence organization.

Introduction

The Army's traditional, or legacy, concept of intelligence operations was designed during the Cold War for conflicts involving centralized and hierarchical state-based threats. It assumed a relatively predictable, doctrinal enemy and a linear, sequential progression of operations through preconflict, conflict, and postconflict operational phases. Despite significant changes in the operational environment in the post–Cold War period (including the emergence of various unconventional and nonstate threats), this concept of intelligence operations was used well into OEF and OIF.

Operational experience and protracted conflicts in Iraq, Afghanistan, and elsewhere have belied the assumption that an intelligence concept of operations and supporting systems designed for Cold War threats could be applicable to the post–Cold War and post-9/11 operational environment. In response to strategic and tactical realities, the Army developed a new operational concept and a new functional concept for intelligence operations. This has resulted in significant changes to how intelligence units, assets, and personnel are organized, structured, and trained and to the ways that intelligence is gathered and used by tactical organizations.

Legacy Organization and Structure

The Army's legacy intelligence organization and structure was designed to detect centralized, hierarchical enemy combat formations and template or predict their actions, movements, and intent based on a likely, linear progression of operations. Once information on enemy movements, actions, and intent was collected, collated, and analyzed— usually at the national, theater, or corps level—it was then transmitted as intelligence down to tactical units preparing for or involved in high-intensity combat operations.

This system of intelligence organization and structure was top-down in nature and predicated on the centralized control and direction of intelligence operations. Accordingly, tactical intelligence assets were sparse, and tactical intelligence staffs were austere. The principal function of legacy intelligence staffs was to receive intelligence from higher headquarters, direct tactical intelligence gathering and processing, and disseminate intelligence products to the commander, staff counterparts, and subordinate units. Tactical intelligence assets were used to provide localized tactical intelligence to battalion, brigade, and division S2 (intelligence) staffs. Although intelligence gathered at tactical echelons did supplement tactical operations and was collated and transmitted to higher headquarters for analysis and dissemination as part of recurring situational reports, unit operations were not typically or principally driven by intelligence gathered at the tactical level. Standard operating procedures and unit organization and staffing were not structured for independent or decentralized unit operations and intelligence gathering.

In legacy intelligence operations, an Army corps was typically supported by an MI brigade, which was further divided into MI battalions (providing direct support to subordinate divisions) and MI companies (providing direct support to maneuver brigades). The MI company provided each maneuver brigade with the following support:

- A commander to help the [brigade] S2 plan.
- C2 for all MI assets (to include any other assets or units attached or supporting the brigade). The MI company command post serves as the control mechanism for all MI assets. Any additional control teams that support the brigade (for example, a tactical HUMINT control team during some types of operations) collocate and operate with the analysis and control team (ACT).
- Analytic support to the S2 and process and fusion systems.
- Organic intelligence collection systems.[1]

Most legacy units did not have robust MI assets that were organic to the organization. Additional MI units and assets, either regionally or functionally based, were assigned or attached in direct support or general support roles (depending on the supported unit's mission requirements or needs). The U.S. Army Intelligence and Security Command (INSCOM) had and (and continues to have) a number of regionally and functionally focused MI units for this purpose.[2]

---

[1] U.S. Army Intelligence Center and Ft. Huachuca, *Combat Commander's Handbook on Intelligence*, Ft. Huachuca, Ariz., Special Text 2-50.4 (Field Manual 34-8), September 2001, p. iii.

[2] The terms *organic, assign, attach, direct support*, and *general support* refer to specific command relationships. In Army Field Manual 101-5-1/Marine Corps Reference Publication 5-2A, a *command relationship* is defined as the "degree of control and responsibility a commander has for forces operating under his command." *Organic* is defined as "[a]ssigned to and forming an essential part of a military organization. Organic parts of a unit are those listed in its table of organization for the Army, Air Force, and Marine Corps, and are assigned to the administrative organizations of the operating forces for the Navy." *Assign* is defined in two parts as follows:

1. To place units or personnel in an organization where such placement is relatively permanent, and/or where such organization controls and administers the units or personnel for the primary function, or the greater portion of the functions, of the unit or personnel

2. To detail individuals to specific duties or functions where such functions are primary and/or relatively permanent.

Changes in the Operational Environment

The operational environments of OEF and OIF (in particular) exposed a number of flaws in the Army's legacy intelligence concept of operations, organization, and structure. The high pace, variety, and decentralized nature of operations conducted in these operational environments resulted in a dearth of actionable intelligence available to tactical commanders.

Increasingly, the intelligence organization, structure, and processes that served the Army well in the combat phase of OIF became less relevant, despite changes and improvements to intelligence organization and structure at the theater level. The operational environment had changed so significantly as to render many of the Army's tactical intelligence capabilities and procedures ineffective:

> The enemy situation was so fluid and local in character that the US Army Intelligence system designed to push down information from division to brigade and then to battalion became increasingly irrelevant. This is not to say that the division G2s and the division-level MI battalions ceased operations. However, their traditional functions and processes were less important than lower-level efforts in the Army's new campaign.[3]

Tactical commanders, by circumstantial necessity and in the absence of relevant intelligence products from higher echelons, were required to generate their own intelligence products in support of decentralized operations:

Attach is the "placement of units or personnel in an organization where such placement is relatively temporary." *Direct support* refers to a "mission requiring a force to support another specific force and authorizing it to answer directly the supported force's request for assistance." *General support* is defined as the "support which is given to the supported force as a whole and not to any particular subdivision thereof" (Headquarters, U.S. Department of the Army, and Headquarters, U.S. Marine Corps, 1997).

[3]   Donald P. Wright and Timothy R. Reese, *On Point II, Transition to the New Campaign: The United States Army in Operation Iraqi Freedom, May 2003–January 2005*, Ft. Leavenworth, Kan.: Combat Studies Institute Press, June 2008, p. 197.

Rather than relying on the standard Cold War era military intelligence (MI) systems and procedures that gathered information at levels above the brigade and then pushed that information down to the tactical level, in Iraq battalion- and even company-size units began conducting their own intelligence operations. This development ran counter to doctrine, and MI professionals expressed concern about the lack of specialized training within the infantry, armor, and other battalions that were busy creating their own intelligence. However, tactical commanders had little choice.[4]

Brigade and battalion commanders increasingly began to conduct intelligence operations with organic maneuver assets, despite their lack of formal intelligence-gathering training:

> This was a major shift in practice. US Army doctrine gave MI Soldiers and units the formal authority to gather, analyze, and disseminate intelligence. The US Army's tactical units, nevertheless, had only a handful of MI Soldiers serving on the staffs of battalions and brigades. The MI officers and noncommissioned officers (NCOs) at these levels did little of their own collection and, other than the armor and infantry battalion S2 sections that could employ organic scout platoons to locate and watch enemy activity, had few assets to do collection. Instead, the Army had designed the MI system to push information from corps and division levels down to brigade and battalion levels where the S2 would make that intelligence relevant for the commander.[5]

Perhaps the greatest challenges facing theater and tactical commanders alike in OIF was an almost total absence of HUMINT capability, little capacity for intelligence integration, and weaknesses in the analytical capacity of intelligence professionals. Protracted COIN and stability operations required different and expanded sets of intelligence competencies, personnel, and analytical methods. In response, the Army reassessed its intelligence capabilities. It made a number of

---

[4] Wright and Reese, 2008, pp. 191–192.

[5] Wright and Reese, 2008, p. 197.

important changes to how intelligence operations were conducted and how intelligence was collected, analyzed, and disseminated by the end of 2003.[6]

Figures B.1 and B.2 show changes to an MI brigade deployed in 2003's task organization.

In addition to the changes to theater-level intelligence organizations, structures, and operations, tactical maneuver units modified their organization for the conduct of intelligence operations. They expanded brigade-level S2 sections, created or expanded S2X (CI and HUMINT) positions, incorporated tactical HUMINT teams into detainee and other operations, and fused and forwarded to subordinate units a range of intelligence products, both organically created and received from higher echelons. SIGINT, HUMINT, imagery intelligence (IMINT, often provided by tactical UAVs), detainee infor-

Figure B.1
An Army Military Intelligence Brigade, Prior to Reorganization in 2003

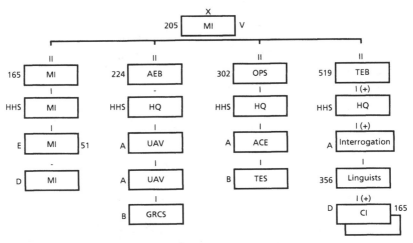

SOURCE: Wright and Reese, 2008, p. 194.
NOTE: AEB = aerial exploitation brigade. GRCS = Guardrail Common Sensor.
TEB = tactical exploitation brigade. TES = tactical exploitation system.
RAND MG1108-B.1

---

[6] The changes included the creation of a fusion center and the expansion HUMINT capabilities in theater.

Figure B.2
An Army Military Intelligence Brigade, Task-Organized, as Deployed in 2003

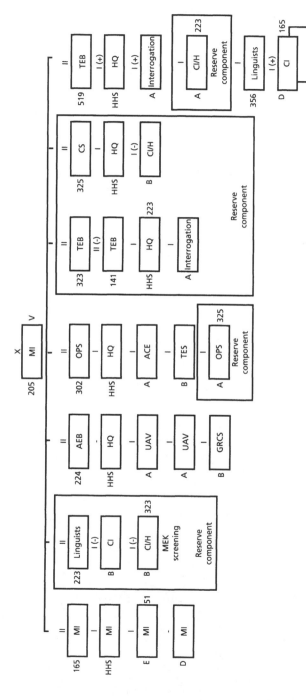

SOURCE: Wright and Reese, 2008, p. 194.
RAND MG1108-B.2

mation, and higher-echelon intelligence were integrated at the tactical level to enable decentralized tactical operations. Many of these short-term changes were permanently incorporated into the organization and structure of future brigade combat teams (BCTs) and battlefield surveillance brigades (BfSBs).

Organizational and Structural Changes in Response to the Operational Environment

In response to changes in the operational environment and the missions that it was assigned to accomplish, the Army developed a new operational concept of full-spectrum operations. This new approach differs from the legacy conception insofar as it no longer assumes that full-spectrum operations are conducted linearly (i.e., through preconflict, conflict, and postconflict stages). Rather, such offense, defense, and stability operations are conducted simultaneously. The Army's new approach to full-spectrum operations also acknowledges that actual and potential threats will be conventional, unconventional, nontraditional, and unpredictable (or any combination thereof) and highly adaptive:

> The future [operational environment] will present the Army with complex and challenging conditions. It will remain difficult to predict and is subject to radical changes and singularities. It may encompass hybrid threats that create multiple dilemmas for our maneuver forces by simultaneous employment of regular and irregular forces, and criminal elements, using an ever-changing variety of conventional and unconventional tactics. Future adversaries will possess weapons of mass destruction (WMD) and technology allowing them to be disruptive over widespread areas.[7]

Operating in this new environment requires the synchronization and integration of available ISR assets and improved management of

---

[7] U.S. Army Training and Doctrine Command, *The U.S. Army Functional Concept for Intelligence 2016–2028*, Fort Monroe, Va., Pamphlet 525-2-1, October 13, 2010, p. 7.

information and knowledge. The resulting approach facilitates intelligence gathering and the sharing of information between and among units and across echelons of command:

> Intelligence, surveillance, and reconnaissance synchronization is the task that accomplishes the following: analyzes information requirements and intelligence gaps; evaluates available assets internal and external to the organization; determines gaps in the use of those assets; recommends intelligence, surveillance, and reconnaissance assets controlled by the organization to collect on the commander's critical information requirements; and submits requests for information for adjacent and higher collection support.[8]

> Intelligence, surveillance, and reconnaissance integration is the task of assigning and controlling a unit's intelligence, surveillance, and reconnaissance assets (in terms of space, time, and purpose) to collect and report information as a concerted and integrated portion of operations plans and orders. This task ensures assignment of the best ISR assets through a deliberate and coordinated effort of the entire staff across all warfighting functions by integrating ISR into the operation.[9]

> Knowledge management is the art of creating, organizing, applying, and transferring knowledge to facilitate situational understanding and decisionmaking. Knowledge management supports improving organizational learning, innovation, and performance. Knowledge management processes ensure that knowledge products and services are relevant, accurate, timely, and useable to commanders and decisionmakers.[10]

To execute these new capabilities and properly organize and train the force for current and future operations, the Army reorganized along

---

[8] Headquarters, U.S. Department of the Army, *Operations*, Washington D.C., Field Manual 3-0, February 27, 2008, p. 7-8.

[9] Headquarters, U.S. Department of the Army, 2008, p. 7-9.

[10] Headquarters, U.S. Department of the Army, 2008, p. 7-10.

six warfighting functions: movement and maneuver, fires, intelligence, sustainment, C2, and protection. The intelligence warfighting function is defined as follows:

> The intelligence warfighting function is the related tasks and systems that facilitate understanding of the operational environment, enemy, terrain, and civil consideration. It includes tasks associated with intelligence, surveillance, and reconnaissance operations and is driven by the commander. Intelligence is more than just collection. It is a continuous process that involves analyzing information from all sources and conducting operations to develop the situation.
>
> Intelligence operations are conducted to provide intelligence in support of all missions. The primary focus of Army intelligence operations is generating tactical intelligence such as, intelligence that supports the planning and conduct of operations. Although the focus is on tactical intelligence, the Army draws on both strategic and operational intelligence resources and, in certain circumstances, conducts intelligence operations at the operational and even strategic level.[11]

The Army's functional concept for intelligence recognizes that the operational environment, and the actual and potential threats contained therein, will require the synchronization, integration, and management of information and intelligence across all the traditional levels of war (strategic, operational, and tactical)—simultaneously and continuously. Intelligence must be gathered and processed at all echelons, and the intelligence function must be both top-down *and* bottom-up in nature.

To accomplish this task, the Army modified its organization and structure for intelligence gathering, analysis, and dissemination by creating BfSBs and by organically incorporating intelligence units, personnel, and assets into each of the BCT variants (heavy, infantry, and Stryker) while maintaining additional regionally and function-

---

[11] U.S. Army Training and Doctrine Command, 2010, p. 10.

ally focused, deployable assets within INSCOM for general and direct support. These units, personnel, and assets allow corps, divisions, and BCTs to process, integrate, and analyze intelligence gathered at higher echelons and to collect tactical intelligence in support of decentralized operations. Furthermore, the Army created a list of functional core competencies for intelligence analysis that will better enable operations in current and future environments.

Battlefield Surveillance Brigades

The BfSB is designed to conduct

> intelligence, surveillance, and reconnaissance (ISR) operations. This capability lets the division commander focus combat power, execute current operations, and prepare for future operations simultaneously. Battlefield surveillance brigades are not designed to conduct guard or cover operations. Those operations may entail fighting to develop the tactical situation; they require a BCT or aviation brigade.[12]

As shown in Figure B.3, the BfSB has organic MI, reconnaissance and surveillance, sustainment, and communication assets.[13]

---

[12] Headquarters, U.S. Department of the Army, 2008, p. C-8.

[13] The figure shows the base BfSB (with the addition of another MI battalion to reflect changes in more recent designs of the BfSB). Additional capabilities, units, equipment, and personnel can be placed under the operational control of or attached to the BfSB, including ground reconnaissance, manned and unmanned Army aviation assets, special operations forces, long-range surveillance, MI assets (HUMINT, aerial exploitation, and national-level assets), and armor, infantry, and combined arms units. The Army defines *operational control* as follows:

> Transferrable command authority that may be exercised by commanders at any echelon at or below the level of combatant command. Operational control may be delegated and is the authority to perform those functions of command over subordinate forces involving organizing and employing commands and forces, assigning tasks, designating objectives, and giving authoritative direction necessary to accomplish the mission. Operational control includes authoritative direction over all aspects of military operations and joint training necessary to accomplish missions assigned to the command. Operational control should be exercised through the commanders of subordinate organizations. Normally this authority is exercised through subordinate joint force commanders and Service and/or functional component commanders. Operational control normally pro-

Figure B.3
Battlefield Surveillance Brigade

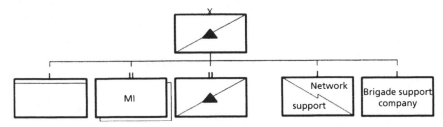

SOURCE: Headquarters, U.S. Department of the Army, 2008, p. C-9.
RAND MG1108-B.3

The subordinate units of the BfSB serve the following functions and provide the following types of support:

- headquarters and headquarters company: C2 of brigade operations
- MI battalion(s): UASs, SIGNINT, and CI/HUMINT
- reconnaissance and surveillance battalion: reconnaissance and surveillance capabilities (including mounted scout platoons and mobile long-range surveillance teams)
- brigade support company: sustainment
- brigade network company: communication architecture (communications through the division area of operations and with Army service component command and national-level intelligence agencies).[14]

Brigade Combat Teams (Heavy, Infantry, Stryker)
The BCT has been restructured to include assets that allow the unit to operate independently and with the structural capacity for aug-

---

vides full authority to organize commands and forces and to employ those forces as the commander in operational control considers necessary to accomplish assigned missions. Operational control does not, in and of itself, include authoritative direction for logistics or matters of administration, discipline, internal organization, or unit training. (Headquarters, U.S. Department of the Army, and Headquarters, U.S. Marine Corps, 1997)

[14] The heavy, infantry, and Stryker BCT functions are outlined in Headquarters, U.S. Department of the Army, 2008, p. C-8.

mentation.[15] Each BCT has an organic reconnaissance, surveillance, and target acquisition (RSTA) capability and an organic MI company (reporting to the brigade S2) that is capable of integrating intelligence produced by higher-echelon commands and agencies and developing tactical-level intelligence products. Each MI company is capable of coordinating intelligence gathering, determining requirements, and conducting short-term intelligence analysis and intelligence preparation of the battlefield but must rely on higher-echelon commands and agencies to provide detailed, long-term analysis. The MI company integrates information and intelligence gathered by other units subordinated to the BCT, including but not limited to counter-battery radar, UAVs, Prophet SIGINT systems, ground surveillance radar, the Remotely Monitored Battlefield Sensor System, tactical HUMINT teams, CI, maneuver unit personnel and S2 sections, and a Common Ground System section that relays imagery and SIGINT products to the BCT.

> As combined arms organizations, BCTs form the basic building block of the Army's tactical formations. They are the principal means of executing engagements. Three standardized BCT designs exist: heavy, infantry, and Stryker. Battalion-sized maneuver, fires, reconnaissance, and sustainment units are organic to BCTs.

> BCTs are modular organizations. They begin as a cohesive combined arms team that can be further task-organized. Commands often augment them for a specific mission with capabilities not organic to the BCT structure. Augmentation might include lift or attack aviation, armor, cannon or rocket artillery, air defense, military police, civil affairs, psychological operations elements, combat engineers, or additional information systems assets. This organizational flexibility allows BCTs to function across the spectrum of conflict.[16]

---

[15] Each of the BCTs has organic ISR assets and personnel, including, but differing slightly in each: HUMINT, all-source intelligence, IMINT, SIGINT, and various reconnaissance, surveillance, and target acquisition resources.

[16] Headquarters, U.S. Department of the Army, 2008, p. C-6.

Figures B.4, B.5, and B.6 show how the three types of BCTs are structured in terms of capabilities and how MI capabilities are integrated at the BCT level. The following descriptions are excerpted from Army Field Manual 3-0, *Operations*.

> Heavy BCTs are balanced combined arms units that execute operations with shock and speed. . . . Their main battle tanks, self-propelled artillery, and fighting vehicle–mounted infantry provide tremendous striking power. Heavy BCTs require significant strategic air- and sealift to deploy and sustain. Their fuel consumption may limit operational reach. However, this is offset by the heavy BCT's unmatched tactical mobility and firepower. Heavy BCTs include organic military intelligence, artillery, signal, engineer, reconnaissance, and sustainment capabilities.[17]

> The infantry BCT requires less strategic lift than other BCTs. . . . When supported with intratheater airlift, infantry BCTs have theaterwide operational reach. The infantry Soldier is the centerpiece of the infantry BCT. Organic antitank, military intelligence, artillery, signal, engineer, reconnaissance, and sustainment elements allow the infantry BCT commander to employ the force in combined arms formations. Infantry BCTs work best for operations in close terrain and densely populated areas. They are easier to sustain than the other BCTs. Selected infantry BCTs include special-purpose capabilities for airborne or air assault operations.[18]

> The Stryker BCT balances combined arms capabilities with significant strategic and intratheater mobility. . . . Designed around the Stryker wheeled armored combat system in several variants, the Stryker BCT has considerable operational reach. It is more deployable than the heavy BCT and has greater tactical mobility, protection, and firepower than the infantry BCT. Stryker BCTs have excellent dismounted capability. The Stryker BCT included military intelligence, signal, engineer, antitank, artillery, recon-

---

[17] Headquarters, U.S. Department of the Army, 2008, p. C-6.

[18] Headquarters, U.S. Department of the Army, 2008, p. C-6.

naissance, and sustainment elements. This design lets Stryker BCTs commit combined arms elements down to company level in urban and other complex terrain against a wide range of opponents.[19]

Figure B.4
Heavy Brigade Combat Team

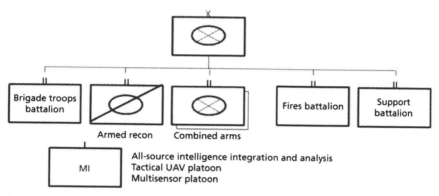

SOURCE: Headquarters, U.S. Department of the Army, 2008, p. C-7.
RAND *MG1108-B.4*

Figure B.5
Infantry Brigade Combat Team

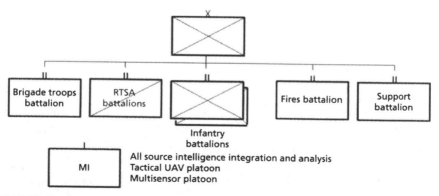

SOURCE: Headquarters, U.S. Department of the Army, 2008, p. C-7.
RAND *MG1108-B.5*

---

[19] Headquarters, U.S. Department of the Army, 2008, p. C-7.

Figure B.6
Stryker Brigade Combat Team

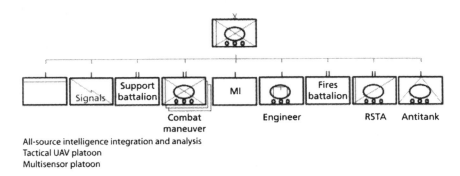

Combat maneuver   Engineer   RSTA   Antitank

All-source intelligence integration and analysis
Tactical UAV platoon
Multisensor platoon

Functional Core Competencies for Intelligence Analysis

The Army's functional core competencies for intelligence analysis fall into four categories: knowledge, skills, characteristics, and abilities.[20] Table B.1 shows the specific components in each category for which intelligence professionals are assessed, trained, and cultivated.

These functional core competencies can be accelerated and refined by training, education, and experience and enhanced by new organizational structures and the habitual training relationships established by these new organizational structures.

Conclusion

With the creation of the BfSB, the expansion of surveillance and reconnaissance assets in the BCTs, the incorporation of organic MI companies in each BCT variant, and the expansion of capabilities in each MI company—including but not limited to all-source intelligence, HUMINT, SIGINT, IMINT, cryptologic specialists, linguists, organic equipment maintainers and integrators, and an inherent fusion

---

[20] U.S. Army Training and Doctrine Command, 2010, p. 68.

Table B.1
Core Competencies for Army Intelligence Analysts

Category	Component
Knowledge	Target knowledge, IC requirements, government plans and policy, customer service, analytic resources
Skills	Critical thinking, literacy, computer literacy, expression, foreign language proficiency, research abilities, information gathering and manipulation, project and process management
Characteristics	Insatiable curious, self-motivated, fascinated by puzzles, voracious observer, voracious reader, fruitfully obsessed, able to take varying perspectives, makes creative connections, playful, has sense of humor, has sense of wonder, able to concentrate intensely, questions convention
Abilities	Communication, teamwork and collaboration, thinking

SOURCE: Adapted from U.S. Army Training and Doctrine Command, 2010, p. 67.

capability—the Army has adapted its legacy organization and structure to conduct full-spectrum intelligence operations in current and future operational environments.

Complete Interview Topics and Questions

Marine Corps Intelligence Organization Informational Interview

Introduction

Hello, my name is _____. I am a researcher from RAND, a private nonprofit research organization with offices in Santa Monica, Washington, D.C., and Pittsburgh. Besides myself, _____, _____, and _____ are participating by phone.

The Director of Marine Corps Intelligence has asked the RAND Corporation to assist in a study of Marine Corps intelligence organization.

The information you provide during this interview will be kept strictly confidential and used for research purposes only. If used, your responses will be aggregated with those of other people, and we will not attribute particular comments or suggestions to specific individuals. Your participation in this interview is voluntary, so if you prefer not to answer a question, or if you want to end this interview for any reason, just let me know.

During this interview, I will be asking about MC intelligence organizational issues and your own experiences with these organizations. I will begin by asking you to describe your current role and responsibilities. After that, we would like to focus the questions on your views about the organizational characteristics of various elements of the Marine Corps intelligence enterprise, including your organization, as well as the broader intelligence-supporting establishment, primarily the Intel Department at HQMC. To the extent that we have time, we would also like to discuss MCIA. Because of this breadth of interest, we may ask you the same question more than once—first referring to your organization and then again with reference to a different part of the enterprise. We also welcome your comments about any aspect of MC intelligence structure and organization from the ground and air combat elements through the command element to the supporting establishment.

Do you have any questions before we start?

You and Your Organization

In this section, we seek to learn a little bit about your background, your relationship to Marine Corps intelligence, and the organization(s) for which you will be providing information.

1. Please tell us a bit about your background and experience in or with Marine Corps intelligence.
2. What is your current position? What are your current roles and responsibilities?
3. We are trying to ascertain organizational details about a number of elements of the Marine Corps intelligence enterprise. Many of the questions that follow will ask about some property of "your organization." This does not need to be your *current* organization if you have significant experience with Marine Corps intelligence in a previous posting. If your current (or a previous) position is in an organization that is a "customer" of Marine Corps intelligence, we'd like your perceptions of MCI organizational characteristics. When we say "your organization," what organization(s) will we be talking about?

Basic Organizational Structure

4. There is some basic information we would like to obtain about the structure of your organization, ideally from an organizational chart:
    a. Whether the organization is structured around divisions, functions, products, customers, geographic regions, processes/workflows, or matrix?

    b. Number of "units" in the organization?

    c. Number of levels, from bottom to top?

    c. Number of levels between lowest- and highest-level personnel?

    d. Lines of authority, responsibility, and accountability?

5. To what extent does the organization consolidate work in a small number of regions of the world or distribute it to many locations? If consolidated, how close to or far from HQ?

6. What written documents or guidance do you use on a regular basis? (For example, a strategic plan, commander's intent, doctrine, or tactics, techniques, and procedures or other formal or informal guidance)

7. What is the 5-10 trend in manpower and budget for your organization?

8. Has there been a recent reorganization or restructuring of the organization? What were the results?

Goals

9. What is the overall mission and vision for your organization, i.e., what is the organization's work activity/what does the organization do?

10. Who are the customers of your organization?

11. When push comes to shove, what takes precedence in the organization—quality, time, or cost? Which two get picked? Which two should get picked?

Relationships with Other Organizations

12. What is your relationship to your customers, e.g., do you collaborate with them? Report to them? Are interactions formal or informal?

13. What is your organization's relationship to other intelligence organizations (e.g., Army intel, CIA, NSA, NGA)?

    a. Are they allies/collaborators or competitors?

    b. What do you get from them? Give to them?

    c. (If they collaborate): Where are they located physically/geographically?

14. Do you have competitors? Who are they? In what ways do you compete?

Complexity and Unpredictability in the Operational Environment

15. What are the top three or four critical factors in your environment that affect the operations and outcomes of your organization? (Probes: DOTMLPF, OPTEMPO, budget, technologies, familiarity with target or region, government relations, political considerations, quality requirements, security requirements.)
16. Choose two factors. How do changes in one affect the other? Are there interdependencies among other factors?
17. For each factor you mention, score its predictability on a scale from 1 (very low) to 5 (very high).

Products

18. What are some examples of your organization's products or services?
    a. Do you think of them as products, services, or something else?
19. To what degree are products or services tailored to individual customers?
20. How do you prioritize between and within customers when there are conflicts in priorities?
21. Business uses the terms "product shelf life" and "product life cycle." Is there an analog in USMC intelligence? What determines the shelf life of your products?

Innovation

22. Do you develop new products or processes? How often?
23. What drives innovation? Is it pulled (customer-driven) or pushed (internally motivated)?
24. To what extent does your organization look to other organizations (intelligence or otherwise) for new products or processes? What organizations do you follow?

25. When faced with changes in the environment, do you find that your organization typically has anticipated these changes or is reacting to them?

26. How quickly can your organization change its processes or products?

27. How much (in scope or degree) can your organization change its processes or products? Is that change incremental or revolutionary?

Internal Coordination

28. Within your organization, to what extent is work done by individuals or by groups?

29. To what degree do units coordinate or collaborate with each other (formally or informally), and in what "order"? (For example, does one unit hand off work to another sequentially; are multiple units involved at one time?)

30. How is this coordination accomplished (e.g., through a centralized office or intermediary, or directly between personnel in the units)?

31. To what extent does your organization find that you have to "reinvent the wheel" or otherwise find out after that fact that you duplicated another group's effort?

32. How difficult would it be for the organization to accommodate major changes into current "rules" or practices?

Distribution of Operations and Decisionmaking

33. Overall, is USMC intelligence organized to ensure centralization of decisionmaking and consistency of work practices across units or autonomy of local units and customization to meet local needs?

34. To what extent are important decisions in your organization made with the larger USMC perspective in mind versus a local perspective?

35. To what extent does the organization decide where to locate its operations based on close proximity to
    a. customers

      b. human resources or assets

      c. suppliers

      d. partners

      e. other resources?

36. In your organization, is it worse (or more common) to miss a good opportunity or to implement a poor solution?

37. Is it better (or more common) to pursue all potential intelligence leads, even if some don't pan out, or to be more selective and pursue only those that are likely to produce intelligence?

Organizational Knowledge, Information Flow, and Information Technology

38. To what extent does the organization rely on internal versus external sources of knowledge to do its work?

39. To what extent is the organization dependent on information supplied by other units or organizations located elsewhere in order to do its work?

40. To what extent are other organizations or units that are dependent on information supplied by your organization located elsewhere?

41. To what extent does the organization rely on IT for analysis?

42. To what extent does the organization rely on IT for communication and collaboration?

43. When/with whom would you use FTF, phone, email, IM, VTC, or digital or physical documents?

44. Do you have the IT infrastructure that you need?

45. To what extent does classification of information impede information flow?

46. Does guardedness or low willingness to share impede information flow within the organization? With external organizations with which you collaborate? If so, how do you get the information you need?

47. Excluding issues of information classification, to what extent could most of the important information that is exchanged within the organization be readily recorded on paper or in a computer system, or is it something "you just know"?

48. To what extent is important information in the organization relatively easy to understand or explain? Are there subtleties to understanding the information, requiring specialized experience or expertise to fully "make sense" of or interpret the information?

Task Design

49. To what extent are tasks or procedures in your organization standardized or variable?
50. To what extent do you execute tasks today much as you did yesterday?
51. Do job descriptions tend to be narrow or broad?

Formalization and Centralization

52. To what extent are subunit decisions and actions directed by HQ or another sole authority versus managed independently by the subunits?
53. To what degree does leadership leave control of decisions, such as budget, operational decisions, and handling work exceptions, to personnel in charge of units or operations?
54. To what extent are there well-known expectations about what is "correct," "acceptable," or "expected" of team members?
55. To what extent are there well-known penalties for violating rules or not meeting expected job behavior?
56. To what extent are team members' actions monitored, recorded, or provided as feedback to either the individuals themselves or leaders?

Process: Competitive Advantage

57. What does your organization provide to your customers that the rest of the intelligence community (outside of USMC) does not or cannot?
    a. What is your competitive advantage?
    b. Where do you need to improve?
58. If your organization were rendered ineffective by a natural disaster or other catastrophic event, what organization might step in

to perform your functions? What would it need to do to take over?

Personnel
59. How many people are in the organization?
60. For your organization, is that sufficient?
61. If not, how many people do you need, and in what capacity?
62. What proportion have an intel or signals MOS as their primary MOS?
63. What proportion are officer, enlisted, civilian, or contractor?
64. What are the average years of experience for each of those categories?
65. Do they have the training and experience you need?
66. To what degree do staff require extensive OTJ training to perform their jobs versus being trained and ready to go upon arrival? (Do you use a "make" or "buy" strategy?)
67. On average, how long does it take for a new team member to perform effectively?
68. How often do individuals rotate in and out of the organization?
69. To what extent is rotation/turnover unanticipated or planned?
70. To what degree are changes in personnel disruptive?
71. To what degree do changes in personnel result in new ideas?
72. To what degree are the MOSs and skills used in this organization valued for promotion and retention in the broader MC?
73. To what extent does performance evaluation in your organization primarily emphasize work process or primarily emphasize results?

Delegation and Uncertainty Avoidance
74. To what extent does leadership maintain control itself or encourage others to take on responsibility for managing work tasks?
75. To what extent does leadership allow direct reports to make important decisions and take action for the organization?
76. To what extent does leadership concern itself with the big picture versus details in decisionmaking?

Organizational Climate and Culture

77. What is the level of trust in the organization, scored from low to high?
78. What is the level of conflict in the organization, scored from low to high?
79. To what extent do people perceive rewards to be fairly distributed across members of the organization, scored from low (very unfair) to high (very fair)?
80. To what degree do people scapegoat or blame others within the organization for problems, scored from low to high? Outside the organization?
81. Which of these best describes your organization?
    a. Personal/like an extended family.
    b. Dynamic and entrepreneurial.
    c. Results-oriented, getting the job done, competitive, and achievement-oriented.
    d. Controlled and structured; governed by formal procedures.

Final Thoughts

Thank you very much for your time and insight.

82. Are there any important organizational issues or concerns that we have *not* asked about? Please elaborate.

Recent History of Marine Corps Intelligence

Intelligence at the End of the Cold War

For the IC, the end of the Cold War ushered in a period of doctrinal and fiscal uncertainty. In an uncertain strategic environment, many observers questioned the need to sustain the intelligence infrastructure built up during the Cold War.[1] In 1994, the chairman of the House committee overseeing intelligence activities asked the Director of Central Intelligence (DCI) how "a changed world has led to a restructuring of requirements" and how that has affected resources and personnel.[2] The questions suggested that, without a Cold War rival, intelligence might not need the resources it needed in the past. David Boren, chairman of the Senate Select Committee on Intelligence and an exceptionally thoughtful participant in conversations about the post–Cold War future for intelligence, was not so sure. He wrote in *Foreign Affairs*, "It is clear that as the world becomes multipolar, more complex and no longer understandable through the prism of Soviet competition, more intelligence—not less—will be needed."[3] A changed strategic context

---

[1] See Gregory F. Treverton, *Reshaping National Intelligence for an Age of Information*, New York: Cambridge University Press, 2001, p. 21.

[2] U.S. House of Representatives, "The Current and Future State of Intelligence," hearing before the Permanent Select Committee on Intelligence, Washington, D.C., February 24, 1994, p. 2.

[3] David L. Boren, "The Intelligence Community: How Crucial?" *Foreign Affairs*, Summer 1992.

meant that intelligence organizations, missions, activities, and products would need to be drastically rethought.

Amid conversations about post–Cold War roles and missions for the IC was a presumption that intelligence, like defense in general, should be less expensive. Expectations for a "peace divided" meant declining defense funding and personnel reductions for many national security institutions. Between 1990 and 1994, Director of National Intelligence (DNI) James Woolsey testified, the intelligence budget declined by 14 percent, and the three major intelligence agencies (NSA, CIA, DIA) were downsizing personnel with a target reduction of 22.5 percent for the decade.[4] "United States' intelligence capabilities are being reduced to a level where," Woolsey warned, "we are skating on thin ice on a warm day."[5]

Desert Storm

An early post–Cold War mission, Operation Desert Storm (1990–1991), offered valuable insight into the capability of existing intelligence organizations to meet new challenges. Despite the operation's combat success, many observers did not like what they saw in terms of intelligence capabilities.[6] Director Woolsey summarized the shortcomings, stating that "commanders found that they had to wait too long for far too few pictures of the battlefield areas. And in the field they had to wait still longer—precious hours during battle—for the pictures to get into their hands because of antiquated procedures for delivering them."[7] The war, to many, demonstrated problems of coordination, technological capabilities, and integration with operations.

---

[4]   U.S. House of Representatives, 1994, p. 21.

[5]   U.S. House of Representatives, 1994, p. 18.

[6]   Much blame was placed on the performance of a new organizational concept for intelligence: the surveillance, reconnaissance, and intelligence groups (SRIGs), established in 1988. The SRIGs were a consolidation of intelligence and communication activities and were intended to serve as an integral part of the MEF. See M. S. Grogan, S. Lima, J. Terando, and G. A. Winterstein, *The Surveillance, Reconnaissance, Intelligence Group Concept and Organization,* Quantico, Va.: Communication Officers School, March 23, 1992.

[7]   U.S. House of Representatives, 1994, p. 18.

The Gulf War experience led the USMC to a stage of significant self-reflection. A series of postwar essays in the *Marine Corps Gazette* pointed out major organizational shortcomings made apparent by the war. To Major Craig Huddleston, a "burning need for tactical intelligence" arose as a major lesson of the conflict.[8] Asserting that marines' requests for CI and interrogation-translation teams had gone unmet, Huddleston called for closer integration of the collection of information and the use of intelligence by the warfighter. "Intelligence guys," Huddleston directed, "take off your trench coats, put on your flak jackets and helmets, and get *down*. We've got a lot to tell you, and we don't know all the questions."[9] Perhaps the most influential commentator on post–Gulf War lessons learned was then-BGen. Paul K. Van Riper. General Van Riper observed operations by I MEF and U.S. Central Command during Operation Desert Storm. In a widely read 1991 article, Van Riper expressed admiration for the USMC's overall performance in the war but identified intelligence as the weakest link.[10] Although he was not the first to comment on shortcomings in USMC intelligence, Van Riper's observations proved so influential as an impetus for reform that his name is still widely associated with subsequent reform as the "Van Riper Plan" (as the 1994 Intelligence Plan is known). In his 1991 article in the *Marine Corps Gazette*, he expressed concern that intelligence marines were too insularly focused on their own craft rather than on getting usable information into the hands of warfighters. "Many seem fascinated with systems and procedures," Van Riper charged, "rather than the product being (or more often not being) provided to the operators."[11] One outcome of this inadequate training was that information was not being adequately analyzed to generate usable intelligence. Van Riper claimed that these shortcomings were "endemic and stem from the way we select, train, and edu-

---

[8]   Craig Huddleston, "Commentary on Desert Shield," *Marine Corps Gazette*, Vol. 75, No. 6, June 1991, p. 33.

[9]   Huddleston, 1991, p. 33.

[10]   Van Riper, 1991.

[11]   Van Riper, 1991, p. 58.

cate our intelligence personnel."[12] Reshaping the USMC intelligence organization to produce and develop personnel who are better suited to effective tactical intelligence would become a central task of the reform agenda that bears his name.

Yet not all observers of the Gulf conflict diagnosed intelligence failures in the same way. Michael H. Decker, who would eventually serve as the USMC's Assistant DIRINT, believed that Van Riper's critique presented an incomplete picture of the problem. He rejected the notion that intelligence officers were smitten by what Van Riper called "systems fascination." Decker attributed critiques such as Van Riper's to "officers who do not understand the capabilities of intelligence collection systems and feel they are being given the run-around when someone tries to explain why a given request can't be fulfilled."[13] Decker suggested that the goal of improving tactical intelligence would have to be a two-way effort: Intelligence marines needed to better understand the needs of the consumers, and operators needed to better understand intelligence activities.

In the same issue of the *Marine Corps Gazette*, Maj. C. E. Colvard echoed Decker's sentiment that improving intelligence would require changes for both operators and intelligence marines. Colvard noted that in Desert Storm, intelligence marines "fought like we trained"; unfortunately, little of that training involved the employment of tactical intelligence. Improving the system for future combat operations would require improvements to the training of intelligence marines to allow them to operate effectively in a battlefield environment. It would also require, he emphasized, taking a "look at how we train our operators to deal with tactical intelligence."[14] Even individuals such as Decker and Colvard, who pushed back against the notion of "intelligence failures" during the Gulf War, argued that there was ample

---

[12] Van Riper, 1991, p. 58.

[13] Michael H. Decker, "Assessing the Intelligence Effort," *Marine Corps Gazette*, Vol. 75, No. 9, September 1991, p. 23.

[14] C. E. Colvard, "Unfortunately, We Fought Like We Trained," *Marine Corps Gazette*, Vol. 75, No. 9, September 1991, p. 21.

room for improvement—not just for intelligence marines but also for the marines who consumed intelligence products.

One area in which both defenders and detractors of the Gulf War performance of USMC intelligence saw the potential for improvement was the training and career development of intelligence professionals. Van Riper, for example, decried the reliance on lateral moves to fill the intelligence billets. He expressed concern that only officers deemed inappropriate for promotion took a tour in intelligence, raising the question about the quality of recruits as well as their fit as analysts. Colvard protested that such attacks on the quality of intelligence marines were unfair and unproductive but admitted that they were unfortunately widespread. "If we want timely, detailed, high-quality tactical information and products, we must develop a Corps-wide attitude change toward intelligence," he asserted. "There seems to be an underlying feeling among officers that a tour in Fleet Marine Force intelligence is a permanent blotch on a career pattern."[15] He called on the USMC to start "providing the command influence and assets to push tactical intelligence into the mainstream."[16]

The 1994 Intelligence Plan (Van Riper Plan)

Conversations about the appropriate lessons to draw from the Gulf War experience led to internal and external reviews of the organization of USMC intelligence.[17] In March 1994, the Commandant of the Marine Corps approved a reform plan. It identified and sought to address six fundamental deficiencies: inadequate doctrinal foundation, no defined career progression for intelligence officers, insufficient tactical intelligence support, insufficient joint manning, insufficient language capability, and inadequate imagery capability.[18] "The Van Riper Plan," as it became known, was the result of an internal review led by General Van Riper and an inspector general review of USMC intelligence per-

---

[15] Colvard, 1991, p. 21.

[16] Colvard, 1991, p. 22.

[17] John W. Johnston, "A Marine Corps Intelligence/Signals intelligence/Electronic Warfare Perspective," *Marine Corps Gazette*, Vol. 79, No. 1, January 1995.

[18] All Marines Memo 100/95, 1995.

formance during the Gulf War.[19] Another major impetus was congressional pressure to improve intelligence performance in the wake of the Gulf War. In 1993, the Senate Committee on Armed Services directed the USMC to submit a "roadmap" for intelligence.[20] Notably, the first roadmap included a provision for the establishment of MCIA in Suitland, Maryland, reflecting the institutional growth of the USMC intelligence organization. Yet, by one account, the original MCIA was organized to interface directly with national intelligence institutions and was "not organized and manned nor permitted by charter to directly assist operational units."[21] MCIA's role in providing reachback capabilities for marines in the field would be developed and expanded in subsequent decades.

The plan outlined seven principles by which the organization was to abide. In 2009, outgoing Assistant DIRINT Michael Decker described these principles as still salient 15 years after they were first articulated: (1) the focus is tactical intelligence; (2) the intelligence focus must be downward; (3) intelligence drives operations; (4) the intelligence effort must be directed and managed by a multidiscipline-trained and experienced intelligence officer; (5) intelligence staffs use intelligence, and intelligence organizations produce intelligence; (6) the intelligence product must be timely and tailored to both the unit and its mission; and (7) the last step in the intelligence cycle is utilization, not dissemination.[22] These principles put USMC intelligence on a path toward prioritizing tactical intelligence and developing intelligence professionals.

In the years that followed, the plan effected change in certain areas and fell short of reformers' expectation in others. A central area that it had targeted for change was personnel policy, training, and pro-

---

[19] Raymond E. Coia, *A Critical Analysis of the I MEF Intelligence Performance in the 1991 Persian Gulf War*, Quantico, Va.: U.S. Marine Corps Command and Staff College, May 22, 1995.

[20] R. J. Buikema, *Integration of Intelligence into Professional Military Education*, thesis, Quantico, Va.: U.S. Marine Corps Command and Staff College, April 18, 1996.

[21] Coia, 1995.

[22] See All Marines Memo 100/95, 1995.

fessional development. Quantitatively, the plan resulted in a 56-percent growth in USMC intelligence manning between 1994 and 2006.[23] This growth reflected an increase in the number of officers from 478 to 975 and an increase in the number of enlisted personnel from 2,642 to 3,893.[24] Rather than relying on lateral moves, the plan established a career track for intelligence marines, allowing a more balanced grade structure. It also established four new entry-level training tracks for officers: ground intelligence, HUMINT, SIGINT, and aviation intelligence.[25] In response to Van Riper's concerns about insufficient doctrinal foundation, by 2001, the USMC had issued numerous doctrine documents on a range of intelligence topics and disciplines.[26] MCIA's role in supporting the operating forces, training and exercises, and subject-matter expertise also grew during this period.[27]

However, observers continued to complain that the reforms were not having the desired impact in many areas. For example, the intent of establishing the training tracks had been to professionalize the force of intelligence marines and allow officers to develop an area of expertise. As recently as 2006, one observer argued that an unintended consequence of the training track "stovepipes" was the creation of "a group of specialists" that hindered the functional integration of the MCISR-E.[28] "We are still organized around discrete Intelligence disciplines and hierarchical echelons," decried the 2010 *MCISR-E Roadmap*, implying that organization is ill suited to meeting future hybrid threats.[29] In the years following the plan's adoption, contributors to the *Marine Corps Gazette* debated areas in which the plan appeared to be succeed-

---

[23] U.S. Marine Corps Intelligence Department, undated, p. 1.

[24] U.S. Marine Corps Intelligence Department, undated, p. 1.

[25] All Marines Memo 100/95, 1995.

[26] Vernie R. Liebl, "The Intelligence Plan: An Update," *Marine Corps Gazette*, Vol. 85, No. 1, January 2001, p. 54.

[27] Liebl, 2001, p. 55.

[28] Matthew Collins, "Beyond the Van Riper Plan: How Are We Growing Intelligence Officers," *Marine Corps Gazette*, Vol. 90, No. 10, October 2006.

[29] U.S. Marine Corps Intelligence Department, undated.

ing and in which its effect had been minimal. One target of reforms that inspired much commentary (and complaint about slow progress) was initiatives to improve the training and selection of intelligence personnel.[30] Another area thought to have seen minimal advancements was the reputation of intelligence personnel. Writers continued to describe a persistent "crisis of credibility" that weakened their efforts to better integrate with operations. In a 2001 review essay of progress to date, Maj. Vernie R. Liebl noted, "The Intelligence Plan gave us a great start," but, he emphasized, much work remained—especially to improve "operationally relevant" intelligence.[31]

Transformation

Meeting post–Cold War challenges required what was popularly termed by the end of the 1990s "transformation" to a military that was equipped to meet threats that were more diffuse, harder to identify, and less tied to a nation-state.[32] Though often vaguely or inconsistently defined, transformation generally referred to DoD's need to undergo large-scale changes in terms of military technology, operating concepts, and military organizations.[33] The transformed military would be more coordinated and employ more innovative approaches to warfare, including ISR.

ISR capabilities were often described in this period as a key element of the transformation agenda. The 2001 Defense Science Board study on transformation included ISR as a key element in achieving information and decision superiority, a capability required for trans-

---

[30] Jeffrey N. Takle, "The Intelligence Plan: A Three-Legged Chair?" *Marine Corps Gazette*, Vol. 86, No. 2, February 2002; Braden W. Hisey, "Producing a More Practical Tactical Intelligence Officer," *Marine Corps Gazette*, Vol. 82, No. 12, December 1998.

[31] Liebl, 2001, p. 57.

[32] See for example, Donald Rumsfeld, "Transforming the Military," *Foreign Affairs*, Vol. 81, No. 3, May–June, 2002.

[33] Ronald O'Rourke, *Defense Transformation: Background and Oversight Issues for Congress*, Washington D.C.: Congressional Research Service, RL32238, November 9, 2006.

formation.[34] The 2001 Quadrennial Defense Review (QDR), which embraced transformation as a major goal, asserted,

> Throughout the Cold War, the singular nature of the strategic threat from the Soviet Union provided U.S. intelligence with a remarkably stable target. Today, intelligence is required to provide political and military leaders with strategic and operational information on an increasingly diverse range of political, military, leadership, and scientific and technological developments worldwide.[35]

In a 2003 report, the Congressional Research Service acknowledged the tight relationship between ISR and transformation efforts: "If ISR does not meet the needs of the 21st century force, much of the effort to shift to new kinds of forces and modes of operation could be wasted."[36]

In this context, which emphasized the centrality of intelligence assets, USMC intelligence underwent significant organizational changes. In 1999, the USMC established three intelligence battalions, one to support each MEF.[37] A year later, the Commandant of the Marine Corps, Gen. James L. Jones, established the I-Dept.[38] Intelligence had been a division of the command, control, communication, computers, and intelligence department. General Jones described the move as both an extension of reforms effected since the Van Riper Plan and a recognition of the vital role of intelligence in the USMC's ability to operate effectively in future strategic environments. In 2001, headquarters raised the profile of MCIA, changing it from a field activity to

---

[34] Defense Science Board, *Transformation Study Report: Transforming Military Operational Capabilities*, Washington, D.C., April 27, 2001.

[35] U.S. Department of Defense, *Quadrennial Defense Review Report*, Washington, D.C., September 30, 2001, p. 38.

[36] Judy G. Chizek, *Military Transformation: Intelligence, Surveillance and Reconnaissance*, Washington, D.C.: Congressional Research Service, RL31425, January 17, 2003, p. 1.

[37] Liebl, 2001, p. 55.

[38] All Marines Memo 021/00, "Establishment of Intelligence Department (Code I) at HQMC," 2000.

a command.[39] The Commandant described the move as "a significant event" that affirmed the "Marine Corps' institutional commitment to improving intelligence support while ensuring that the Marine Corps Intelligence Activity remains a true 'center of excellence.'"[40]

Issued in draft form in 2006, the *Marine Corps 2005–2015 ISR Roadmap* described the period since the Intelligence Plan as the "first intelligence transformation period."[41] The name captured the organizational, relational, and technological changes effected by both the Intelligence Plan and enthusiasm for "transformation." The document described the USMC as an institution that had undergone significant organizational and technological change to improve interoperability and coordination.

Reforming National Intelligence Institutions

National intelligence institutions have grown and changed considerably since the end of the Cold War. Reform efforts, accelerating since September 11, 2001, have particularly emphasized improvements to the coordination of intelligence activities. Yet, centralization is neither easy to achieve nor a goal that is universally shared. Indeed, many observers see the redundancies in the system as a key element of effective intelligence.[42]

One major area of concern for reformers interested in improved coordination among national intelligence institutions has been the question of appropriate leadership of the IC. For decades the DCI was simultaneously head of the 16-agency IC (of which the USMC is a member) and director of the CIA. The arrangement had long struck

---

[39] Robert W. Livingston, "Marine Corps Intelligence Activity—Excellence in Expeditionary Intelligence," *Marine Corps Gazette*, Vol. 79, No. 4, April 1995.

[40] Marine Administrative Message 079/01, "Command Activation," February 2001.

[41] Headquarters, U.S. Marine Corps, *Marine Corps 2005–2015 ISR Roadmap*, draft, October 2, 2006.

[42] Mark M. Lowenthal, *Intelligence: From Secrets to Policy*, Washington, D.C., Congressional Quarterly Press, 2006.

reformers as insufficient for coordinating the programs and resources of the sprawling, heterogeneous membership of the IC. Soon after the end of the Cold War, Senator David Boren and Representative David McCurdy, respective chairmen of the Senate Select Committee on Intelligence and the House Permanent Select Committee on Intelligence, introduced what was ultimately unsuccessful legislation to establish the position of DNI to coordinate intelligence programs and resources.[43] The move would have centralized authority to a greater degree, providing the new director with an institutional location better suited to overseeing and coordinating intelligence activities than the DCI. Senator Boren wrote, "It is necessary to give one person the power to coordinate and set priorities for the entire intelligence community."[44] The package of reforms was not adopted, partly due to strong opposition from institutions with a stake in intelligence reform: the DoD and the congressional armed services committees.[45] The reform effort (and many like it before and since) highlights the complex organizational networks governing IC priorities, practice, resources, and oversight. Effecting change in intelligence organizations requires careful negotiation of complex incentive systems, disciplinary boundaries, and oversight responsibilities.

The events of September 11, 2001, perceived intelligence shortcomings in the lead-up to the Iraq War, and the 2004 release of the findings of the 9/11 Commission proved to be a powerful impetus for major intelligence reform. The 9/11 Commission renewed calls to reform the leadership of the IC. As Boren and McCurdy had a decade earlier, the commission recommended installing a DNI.[46] Commission members found that the DCI as both head of the CIA and coordinator of the IC had "too many jobs" to provide effective leadership and

---

[43] Alfred Cumming, *The Position of Director of National Intelligence: Issues for Congress*, Washington, D.C.: Congressional Research Service, RL32506, August 12, 2004.

[44] Boren, 1992.

[45] Richard A. Best, Jr., *Proposals for Intelligence Reorganization: 1949–2004*, Washington, D.C.: Congressional Research Service, RL32500, September 24, 2004a.

[46] Thomas J. Nicola, *9/11 Commission Recommendations: Intelligence Budget*, Washington, D.C.: Congressional Research Service, September 27, 2004.

had insufficient control over IC resources to oversee and coordinate IC activities.[47] "The DCI has to direct agencies without controlling them," said the commission's report. "He does not receive an appropriation for their activities, and therefore does not control their purse strings. He has little insight into how they spend their resources."[48] In December 2004 the Intelligence Reform and Terrorism Prevention Act (P.L. 108-458) was signed into law, codifying the recommendation to establish a DNI.[49] In 2005, the Senate confirmed John Negroponte as the first DNI. The establishment of the office signifies the move—since the end of the Cold War and especially since 9/11—toward increased centralization and coordination of intelligence activities. As a result, the DIRINT, a member of the IC, has both more consumers for USMC intelligence products and more institutions with which to integrate.

Another area of concern among recent reformers of national intelligence institutions has been the intelligence budget. Between 1995 and 2004, the U.S. intelligence budget was divided into three components: the National Intelligence Program (NIP), which supported all foreign intelligence and CI activities; the Joint Military Intelligence Program (JMIP), which supported all defensewide intelligence requirements; and the Tactical Intelligence and Related Activities (TIARA), which supported the aggregation of funding for tactical MI managed by the individual services.[50]

Notably, TIARA support for tactical MI programs was managed by the individual military services, a decentralized structure that the USMC Assistant DIRINT reportedly found useful. Michael Decker testified that ISR capabilities "should remain in TIARA so the commander will have an ownership stake in not only making them part of

---

[47] Nicola, 2004.

[48] Quoted in Richard A. Best, Jr., *Intelligence Community Reorganization: Potential Effects on DoD Intelligence Agencies*, Washington, D.C.: Congressional Research Service, RL32515, December 21, 2004b.

[49] Richard A. Best, Jr., *Intelligence Issues for Congress*, Washington, D.C.: Congressional Research Service, September 18, 2009.

[50] Stephen Daggett, *The U.S. Intelligence Budget: A Basic Overview*, Washington, D.C.: Congressional Research Service, RL21945, September 24, 2004.

his team in combat, but in preserving and enhancing these capabilities during Service planning, programming, and budgeting."[51] USMC ISR programs were supported by the JMIP and TIARA budgets.[52] This structure said more about the consumers of intelligence products than about the organization of intelligence activities. Under this structure, NSA cryptologic activities could fall under either NIP or JMIP, depending on who the ultimate consumer was to be.[53] The structure of the intelligence budget also had implications for the oversight of intelligence activities. While the Senate Intelligence Committee exercised oversight of NIP, it had little authority over TIARA. The House Permanent Select Committee on Intelligence shared responsibility for oversight of the TIARA budget with the House Armed Services Committee.[54]

The structure of the intelligence budget became a key area for intelligence reform efforts. Advocates for reform have argued that the categorization of intelligence expenditures no longer reflects the strategic environment in which intelligence organizations operate. According to a 2005 Congressional Research Service report, "Over a number of years it has become apparent that, to consumers of intelligence, distinctions among NIP, JMIP, and TIARA programs are becoming indistinct."[55] The report noted that both current combat operations and projections about future strategic environments have blurred the line between tactical intelligence and MI.[56] In September 2005, calls to align the structure of the intelligence budget with the nature of intelligence challenges came to fruition, and JMIP and TIARA activities

[51] U.S. Senate, "Marine Corps Intelligence Programs and Lessons Learned in Recent Military Operations," hearing before the Committee on Armed Services, Subcommittee on Strategic Forces, Washington, D.C., April 7, 2004.

[52] U.S. Senate, 2004.

[53] Daggett, 2004.

[54] Frank J. Smist, Jr., *Congress Oversees the United States Intelligence Community, 1947–1994*, Knoxville, Tenn.: University of Tennessee Press, 1994.

[55] Richard A. Best, Jr., *Intelligence, Surveillance, and Reconnaissance (ISR) Programs: Issues for Congress*, Washington D.C.: Congressional Research Service, RL32508, February 22, 2005.

[56] Daggett, 2004.

were merged into a new category, the Military Intelligence Program (MIP). This change yielded to a two-part intelligence budget: the NIP and the new MIP. The NIP now consisted of programs that supported national decisionmakers and was overseen by the DNI. Oversight of the MIP became the purview of the Under Secretary of Defense for Intelligence, itself a new position. Established in 2003 in response to a perceived need to coordinate DoD's intelligence, intelligence policy, plans, programs, and resources, and the Under Secretary of Defense for Intelligence became a point of contact between DoD and the leadership of the IC.[57] Today, within this new budget structure, the MIP supports 92 percent of the USMC intelligence budget, and the NIP supports just 8 percent of activities.[58] DIRINT is the USMC manager of the MIP and thus responsible for coordinating USMC ISR programs with national and defense intelligence organizations.[59]

National intelligence institutions have undergone significant changes in the post–Cold War and, especially, the post-9/11 era. In general, reforms have increased the centralization of intelligence funding decisions. Changes such as the establishment of the DNI and the folding of tactical intelligence activities, previously managed by the services, into a military intelligence budget managed by the Under Secretary of Defense for Intelligence reflect a move toward increased centralization. At the same time, however, it is important to note that effecting change in intelligence organizations is exceptionally difficult. The members of the IC represent strong institutions, governed by complex and intertwining oversight of programs and budgets. For USMC intelligence, centralizing and coordinating reforms may place increased emphasis on relationships with the IC. Yet, the uneven history of organizational change also points to the difficulty of coordinating so many different organizations.

---

[57] Best, 2004b.

[58] U.S. Marine Corps Intelligence Department, 2010.

[59] Marine Corps Order 3900.15B, "Marine Corps Expeditionary Force Development System (EFDS)," March 10, 2008, p. 14.

USMC Intelligence Goes to War

For a decade, the USMC intelligence organization has been at war. The extended period of combat operations has posed significant challenges, but it has also offered unique opportunities. One opportunity that the experience has afforded is the development of an exceptionally experienced generation of intelligence marines—what one observer called "the most experienced group of intelligence professionals in history."[60] Preserving this expertise will be an important challenge as combat operations end. To meet future challenges, the USMC has pursued innovative approaches to organizing its intelligence resources and activities to adapt to an operating environment characterized by irregular threats from state and nonstate actors. One study found that USMC intelligence had adapted well to meeting irregular threats in Iraq and Afghanistan; the author perceived the success to be more in spite of than because of intelligence doctrine, training, and organization.[61] A widely discussed innovation was the distribution of intelligence below the battalion level. Company-level intelligence reflected a bottom-up innovation to operate in a dynamic environment of irregular threats.

Evolving Intelligence Roles

As the nature of combat operations in Iraq and Afghanistan has evolved over a decade of war, the role of intelligence in support of the conflicts has been the subject of significant discussion. From conventional "forced-entry" operations to counterterrorism and COIN operations (in two very different physical, political, economic, and cultural environments), marines and their institutions have been asked to be agile and flexible. Across this range of operations, intelligence has played a key and evolving role. It has been employed, for example, in conventional operations and in the identification and targeting of terrorists and insurgents.

---

[60] John M. Wear, "Educating Intelligence Specialists," *Marine Corps Gazette*, Vol. 93, No. 5, May 2009, p. 48.

[61] Matthew A. Reiley, *Transforming USMC Intelligence to Address Irregular Warfare*, thesis, Quantico, Va.: U.S. Marine Corps Command and Staff College, 2008, p. 1.

Many of these activities have been supported by unmanned systems, demand for which boomed during the wars in Iraq and Afghanistan.[62] Data from unmanned systems has yielded intelligence that, according to a Defense Science Board report, has "proven invaluable to both national decision makers and to battlefield commanders."[63] UAV platforms, both in theater and at the tactical level, support an increasing number of sensors to meet mounting demands for information. Marines have relied on small, unmanned drones, such as the Raven and the Wasp, often leveraged from other services to meet their ISR needs for the war.[64] In the FY 2011 budget, the USMC requested funding for its own system: the Small Tactical Unmanned Air System. This move reflects the prominent role that ISR technologies are anticipated to play in future conflicts.

Yet, to some observers, evolving roles for intelligence organizations should not focus solely on opportunities afforded by new technologies. In January 2010, a widely read report coauthored by the senior intelligence officer in Afghanistan, MG Michael T. Flynn, advocated a more human and less technological basis for executing effective COIN operations.[65] To meet the challenges of operating in Afghanistan, General Flynn and his coauthors pushed for fundamental change to the intelligence organization, activities, and products. *Fixing Intel* argued that the intelligence community in Afghanistan was fixated too much on battling insurgents and too little on understanding the political, economic, and cultural context in which it operated. This myopic focus on the enemy rather than the "environment that supports it" relegated intelligence institutions to being reactive rather than proactive.[66] Addressing this shortcoming, the authors argued, called for the

---

[62] James W. McMains, "The Marine Corps Robotics Revolution," *Marine Corps Gazette,* Vol. 88, No. 1, January 2004.

[63] Joint Defense Science Board Intelligence Science Board Task Force, *Integrating Sensor-Collected Intelligence,* Washington, D.C., November 2008.

[64] Daniel P. Taylor, "Eyes in the Sky: Unmanned Aerial Vehicles Expand Marine Corp ISR Capabilities," *Seapower,* April 2010, p. 12.

[65] Flynn, Pottinger, and Batchelor, 2010a.

[66] Flynn, Pottinger, and Batchelor, 2010a, pp. 7–8.

production of intelligence products that communicated more nuance, context, and content. "Microsoft Word, more than PowerPoint," Flynn and his co-authors argued, "should be the tool of choice for intelligence professionals in COIN."[67]

The dynamic and highly contingent combat environment led many observers to emphasize the importance of effective integration and coordination of intelligence activities. Notably, such conversations bore many similarities to contemporary conversations about the need to improve integration of national intelligence institutions. One captain writing in the *Marine Corps Gazette* called on intelligence analysts to change the way they viewed their disciplines relative to others: "It is easy to understand how these disciplines can travel along nonintersecting, parallel lanes in search of the same end because the design of intelligence units has compartmentalized specialties."[68] This would no longer suffice, he argued. Rather, analysts must cultivate a genuine proficiency and interest in other disciplines.

Evolving Roles for Marines

In 2006, the Marine Corps *ISR Roadmap* called the second transformational period for the intelligence organization "Expeditionary Maneuver Warfare (EMW) Transformation."[69] EMW, the "capstone warfighting concept for the 21st century," aimed to develop "strategically agile and tactically flexible MAGTFs with the operational reach to project relevant and effective power across the depth of the battlespace." USMC intelligence was designed to support EMW by providing commanders with "all-source, fused intelligence," with speed and agility.[70]

---

[67] Michael T. Flynn, Matt Pottinger, and Paul D. Batchelor, "Fixing Intel in Afghanistan," *Marine Corps Gazette*, Vol. 94, No. 4, April 2010b, p. 67.

[68] William E. DeLeal, "Finding a Needle in a Stack of Needles," *Marine Corps Gazette*, Vol. 93, No. 1, January 2009, p. 30.

[69] As discussed earlier, the first transformation period (1994–2005) was ushered in by the Intelligence Plan.

[70] Headquarters, U.S. Marine Corps, 2006, p. 7.

The impetus for the intelligence transformation beginning in 1994 had been the perceived shortcomings of USMC intelligence capabilities during the Gulf War. Thus, the impetus for the EMW transformation reflected the lessons of an organization that had drawn lessons learned from a decade of combat operations. The Commandant testified in February 2010 that the war highlighted the need to get intelligence products into the hands of warfighters. The MCISR-E supported MAGTF warfare, he explained, by organizing "all of the intelligence disciplines, sensors, and equipment and communication architecture into a single capability that is integrated and networked across all echelons."[71]

Company-Level Intelligence Cell and "Intel at the Grassroots"

Intimately linked to the marine concept of expeditionary maneuver warfare was a widely noted USMC intelligence innovation to allocate intelligence resources to the company level. In an abridged version of *Fixing Intel* published in the *Marine Corps Gazette*, Flynn and his coauthors pointed to recent innovations in USMC intelligence as a model for desired organizational change.[72] In a section headed "Intel at the Grassroots," the authors highlighted the USMC's efforts to distribute intelligence resources down to the company level. CLICs have been a major innovation of USMC intelligence since it went to war in 2001. It represents both the integral role played by tactical intelligence in combat operations and an innovative response to the needs of marines on the ground.

Flynn and his coauthors attributed the turnaround in Nawa, a largely agricultural district in Helmand Province, Afghanistan, to organizational innovations in USMC intelligence. They said that the district had proved to be a challenging insurgent stronghold until July 2009, when the USMC shifted to COIN techniques that focused on understanding the environment, not just the enemy.[73] With intelli-

---

[71] U.S. House of Representatives, "2010 Posture of the United States Marine Corps," hearing before the Committee on Armed Services, Washington, D.C., February 24, 2010, p. 4.

[72] Flynn, Pottinger, and Batchelor, 2010b.

[73] Flynn, Pottinger, and Batchelor, 2010b, p. 64.

gence analysts at the company level, the marines "armed themselves with a network of human sensors who could debrief patrols, observe key personalities and terrain across the district, and—crucially—write down their findings."[74] Flynn and his coauthors linked these efforts to distribute intelligence down to the company level with the larger goal of effecting sweeping changes to intelligence activities for COIN. Similarly, another set of observers linked the CLICs with a realization that COIN is fundamentally a *political activity*.[75] Company-level intelligence offered the promise of equipping commanders with key information about the political environment in which they operated.

The concept of the CLIC emerged from conversations about improving training, manning, and equipping of platoons and squads.[76] Distributed operations (as the concept was known until the term *enhanced company operations* won favor) aimed to align USMC resources to required capabilities during the conflicts in Iraq and Afghanistan. Col. Vincent J. Goulding, Jr., traced company-level intelligence back to the June 2007 Irregular Warfare Conference, at which the idea of formalizing the CLIC arose as a lively topic of discussion.[77] According to Goulding, intelligence activities at the company level were not new, but up to that point, "company commanders were creating this capability ad hoc and out of hide."[78] Ad hoc efforts reflected an increased reliance on the company commander for battlefield functions and the USMC-specific "ethos of maneuver warfare predicated on intelligence-driven operations."[79]

The bottom-up efforts to expand intelligence capabilities at the company level were formalized by the Commandant of the Marine Corps in "A Concept for Enhanced Company Operations," published

---

[74] Flynn, Pottinger, and Batchelor, 2010b, p. 64.

[75] Morgan G. Mann and Michael Driscoll, "Thoughts Regarding the Company-Level Intelligence Cell," *Marine Corps Gazette*, Vol. 93, No. 6, June 2009, p. 28.

[76] Platoons make up companies; companies make up battalions.

[77] Vincent J. Goulding, Jr., "Enhanced Company Operations," *Marine Corps Gazette*, Vol. 92, No. 8, August 2008, p. 17.

[78] Goulding, 2008, p. 17.

[79] Goulding, 2008, p. 18.

by the *Marine Corps Gazette* in August 2008: "Conventional wisdom tells us that the battalion is the smallest tactical formation capable of sustained independent operations," but, General Conway noted, "current operations tell us it is the company."[80] He placed special emphasis on *intelligence* as a means of developing enhanced company operations capabilities. The company commander must "collect, assess, and distribute actionable intelligence, up, down, and across."[81] Company-level intelligence required intelligence activities to improve situational awareness, collection and production of timely and accurate intelligence, collection management, and intelligence management.

While the CLIC has been described as an important organizational innovation, some worry that the bottom-up change lacks sufficient support from above. One observer, Capt. Edward P. Graham, recently argued that the USMC "is not adequately supporting the company-level intelligence cell (CLIC) concept that is proven to be effective in an asymmetric fight."[82] He asserted that CLIC marines lacked sufficient training to perform the broad range of functions they were called on to perform. To Graham, despite the long experience of war, "tactical intelligence training" continued to take "a back seat in initial training to conventional intelligence collection and processes."[83] In this reading, top-down USMC institutions for training and supporting intelligence activities failed to keep pace with bottom-up innovations in warfighting.

Personnel Issues

The long war has cast a heavy burden on the USMC. One account appearing in the *Marine Corps Gazette* found evidence of repeated and lengthy deployments among USMC intelligence personnel. "Retention," the author said, "has become enough of a problem in the com-

---

[80] James T. Conway, "A Concept for Enhanced Company Operations," *Marine Corps Gazette*, Vol. 92, No. 12, December 2008, p. 59.

[81] Conway, 2008, p. 59.

[82] Edward P. Graham, "Company-Level Intelligence Cell," *Marine Corps Gazette*, Vol. 94, No. 3, March 2010, p. 20.

[83] Graham, 2010, p. 22.

munity that the Marine Corps has once again had to resort to forced lateral moves to staff its intelligence field."[84] In 2006, in response to such challenges, the Commandant of the Marine Corps announced a plan to grow the force. His initiative to grow active-duty end strength to 202,000 marines, "202K," was intended to achieve dwell ratios desired by the Secretary of Defense. The "202K" increase implied a growth from 5,122 intelligence marines to 6,222, intended to be complete by early 2011.[85]

In addition to numbers, the long war has also highlighted the need to bolster key areas of expertise. The war experience has, for example, made the USMC a noted leader in cultural intelligence.[86] A report from the U.S. Marine Corps Center for Lessons Learned noted in 2008 that CI/HUMINT exploitation teams proved instrumental in providing actionable intelligence in Iraq.[87] Yet marines reported persistent understaffing of linguists and women on the teams. Without female HUMINT marines, access to intelligence from female civilians proved to be a challenge. The 2006 *ISR Roadmap* echoed the concern about a lack of trained linguists. It noted that the USMC had made a concerted effort to streamline the development and acquisition of trained linguists for operations in the global war on terrorism, but there was room for improvement. One result of efforts to improve the preparation of intelligence marines in key areas of expertise was the establishment of the U.S. Marine Corps Intelligence Schools Command to coordinate the training and education needs for intelligence.[88]

The wartime experience signaled to many observers that training analysts for future conflicts is a broader issue than acquiring narrow new areas of expertise. Overcoming myriad hurdles to effective integra-

---

[84] Collins, 2006.

[85] U.S. Marine Corps Intelligence Department, 2010, p. 1.

[86] James L. Higgins, Michelle L. Trusso, and Alfred B. Connable, "Marine Corps Intelligence," *Marine Corps Gazette*, Vol. 89, No. 12, December 2005; James W. Lively, "Cultural Education," *Marine Corps Gazette*, Vol. 91, No. 4, April 2007.

[87] U.S. Marine Corps Center for Lessons Learned, *Counterintelligence/Human Intelligence Exploitation Operations: Quick Look Report*, Washington, D.C., 2008, p. 2.

[88] Headquarters, U.S. Marine Corps, 2006, p. 17.

tion of intelligence is going to require technological means of facilitating information sharing, Flynn and his co-authors wrote, but it is also going to require changes to the selection and training of analysts.[89] Intelligence institutions needed analysts "empowered to methodically identify everyone who collects valuable information, visit them in the field, build mutually beneficial relationships with them, and bring back information to share with everyone who needs it."[90] This prescription represented a holistic approach to the gathering of information and distribution of intelligence products. The target for study by the new breed of analysts was not just the enemy but everyone with valuable information, and the most immediate goal was not attacking the enemy but *building relationships* with the population.[91] LTC Morgan G. Mann and Capt. Michael Driscoll argued that such a diverse skill set was particularly important in CLICs: "Successful CLICs possess attributes that include analytical capability, prior operational deployment, language training, and computer skills," as well as "curiosity, 'street smarts' and effective written and oral communication."[92]

Supplemental Funding for ISR Activities

The wars in Iraq and Afghanistan have been funded through supplemental appropriations, with a significant but undisclosed amount going to intelligence-related activities.[93] While the supplemental process is intended to provide warfighters with the flexibility necessary to operate in highly contingent combat situations, the extent of its use and the nature of activities it has come to support have been concerns in Congress. Growing war expenditures for ISR activities have made Congress particularly sensitive to supplemental spending. The House Permanent Select Committee for Intelligence reported in 2002,

---

[89] Flynn, Pottinger, and Batchelor, 2010b.

[90] Flynn, Pottinger, and Batchelor, 2010b, p. 66.

[91] Joseph Davidoski, "More Than Mapmakers," *Marine Corps Gazette*, Vol. 94, No. 9, September 2010.

[92] Mann and Driscoll, 2009, p. 28.

[93] Best, 2005, p. 7.

The "advantage" of the supplemental process to the Intelligence Community is that pressing budgetary demands can be met in a shorter time (and with fewer bureaucratic hurdles) than the regular yearly process. However, by continuing to rely on supplemental appropriations year after year, the Intelligence Community risks fostering a budget process that is ripe for abuse and long-term funding gaps.[94]

The House Intelligence Committee repeatedly expressed frustration that supplementals for ISR were increasingly used to bypass base-year programming. A 2004 Senate Intelligence Committee report stated, "While the practice of funding baseline expenditures using supplemental vehicles has become more prevalent in the past 10 years . . . it is time to rein in this practice."[95] The reliance on supplementals for ISR has allowed broad investment in wartime capabilities, but it has also negatively affected the financial stability of some programs.

War funding for USMC intelligence activities has offered both risks and opportunities. The 2010 *MCISR-E Roadmap* noted, "The Long War, and the MIP contribution in OIF and OEF in particular, has been sustained by additional funding through Supplemental and Overseas Contingency funds."[96] In FY 2009 supplemental spending for USMC intelligence amounted to $120 million.[97] While these funding streams offered opportunities, reliance on supplementals carried risks as well. As the roadmap predicted, "It is likely that growing fiscal austerity will place greater pressure on . . . the DoD budget"; such "funding is therefore likely to shrink significantly."[98] Anticipating a

---

[94] U.S. House of Representatives, Permanent Select Committee on Intelligence, *Intelligence Authorization Act for Fiscal Year 2003*, H.Rept. 107-592, July 18, 2002, p. 15, quoted in Best, 2005, p. 8.

[95] U.S. Senate, Select Committee on Intelligence, *To Authorize Appropriations for Fiscal Year 2005 for Intelligence and Intelligence-Related Activities of the United States Government, the Intelligence Community Management Account, and the Central Intelligence Agency Retirement and Disability System*, S.Rept. 108-258, May 5, 2004, p. 10, quoted in Best, 2005, p. 9.

[96] U.S. Marine Corps Intelligence Department, 2010, p. 21.

[97] U.S. Marine Corps Intelligence Department, 2010, p. 21.

[98] U.S. Marine Corps Intelligence Department, 2010, p. 21.

period of fiscal austerity has provided an impetus for the MCISR-E to seek efficiencies through technological and organizational innovation.

Current Guidance Regarding the Strategic Environment

National and DoD-Level Guidance

National Security Strategy (2010)

The current National Security Strategy (NSS), signed by President Barack Obama in May 2010, articulates a characteristic balance between pragmatism and idealism: a strategy for both "the world as it is" and for realizing "the world we seek."[1] It describes "the world as it is" as a strategic environment of nonstate actors, violent extremists, the threat of unsecured nuclear materials, cyber-threats, and terrorists that threaten Americans at home and U.S. interests abroad. It also describes the challenges of a catastrophic economic recession and churning global demographic, resource, and economic trends. To address these challenges, the NSS calls for a strategy that both directly addresses the world in the near term and lays the foundation at home for "the horizon beyond" current conflicts. To this end, the NSS emphasizes a way toward "the world we seek": "a world in which America is stronger, more secure, and is able to overcome our challenges while appealing to the aspirations of people around the world."[2]

For the country to operate in the "world as it is," the strategy calls for employing both military and diplomatic tools. It calls for supporting sovereign governments in Iraq and Afghanistan and working to ensure the physical security of those populations. It also calls for active

---

[1] Office of the President of the United States, *National Security Strategy*, Washington, D.C., May 2010.

[2] Introductory remarks by President Obama on release of the NSS.

engagement with the international community and international institutions. Realizing "the world we seek," it argues, would require aligning national strategy with four national goals: (1) security of the United States, its citizens, and its partners; (2) prosperity from a strong, innovative, and growing U.S. economy in an open international system; (3) respect for universal values at home and abroad; and (4) an international order, with leadership provided by the United States, that promotes peace, security, and opportunity through stronger cooperation.

Notably, the NSS conflates issues generally thought of as domestic issues with those considered national security issues. In the strategy, economic institutions, energy security, education and global competitiveness, and the national deficit are part of a single story alongside threats from nuclear powers, nonstate actors, and violent extremists. "What takes place within our borders," the document states, "will determine our strength and influence beyond them."[3] Shoring up the health of the nation's domestic energy portfolio, educational institutions, and fiscal future would be key foundations for increasing the nation's prosperity and role in the international community. The NSS describes itself as a whole-of-government approach, in which the boundaries between government organizations need to become more fluid than ever to solve common problems.

In line with this holistic approach, the NSS describes intelligence as a key asset for addressing threats at home and abroad:

> Our country's safety and prosperity depend on the quality of the intelligence we collect and the analysis we produce, our ability to evaluate and share this information in a timely manner, and our ability to counter intelligence threats. This is as true for the strategic intelligence that informs executive decisions as it is for intelligence support to homeland security, state, local, and tribal governments, our troops, and critical national missions.[4]

---

3 Office of the President of the United States, 2010, p. 2.

4 Office of the President of the United States, 2010, pp. 15–16.

Improving the sharing of intelligence products across U.S. government institutions (from homeland security to national security), and between the United States and its allies, offers a means of making governmental boundaries more fluid in pursuit of solutions to common problems.

National Defense Strategy (2008)

The National Defense Strategy (NDS), released by the Secretary of Defense in 2008, builds on the lessons learned from past operations and strategic reviews.[5] The NDS is informed by the President's NSS and, in turn, informs the National Military Strategy (NMS). It provides a framework for other DoD strategic guidance, such as campaign and contingency planning, force development, and intelligence. The NDS states that, in the immediate future, the strategic environment will be defined by a global struggle against a violent extremist ideology seeking to overturn the international system. It also identifies the threat of irregular conflicts against insurgents and other nonstate actors, rogue states pursuing nuclear capabilities, and the rising military power of nation-states, such as China.

The NDS points out that DoD needs to plan for operations in future security environments that will be shaped by the interaction of powerful strategic trends. It projects that, over the next 20 years, global trends in such areas as demographics, the distribution of resources, access to energy sources, and climatic and environmental change will combine to create a context of churning social, cultural, and technological change. This dynamic context creates complexities that the NDS must take into account.

Both strategic threats and the global context informed the five key objectives for DoD outlined in the NDS. First, DoD must defend the homeland against state and nonstate actors and against devastating effects of national emergencies. Second, it must win the long war against violent extremism and prevail in irregular campaigns, such as operations in Iraq and Afghanistan. Third, it must promote regional and international security. Fourth, it must maintain the forces nec-

---

[5] U.S. Department of Defense, *National Defense Strategy*, Washington, D.C., June 2008.

essary to deter conflict or dissuade a range of potential adversaries. Finally, DoD must win the nation's wars.

Achieving these ends will involve organizational, technological, and diplomatic changes. Notably, one important tool cited in the strategy is human and technological support for intelligence activities; "DoD is pursuing improved intelligence capabilities across the spectrum." Organizationally, the document recommends improved integration and coordination of DoD components.

National Military Strategy (2004)

In the NMS, the Joint Chiefs of Staff translate the White House's vision in the NSS into an implementation of the Secretary of Defense's NDS and then into courses of action for the armed forces.[6] The intent of the document is to derive objectives, missions, and requirements from an analysis of the NSS, the NDS, and the Joint Chiefs' understanding of the strategic context. The version released in 2004 prioritized the threat of international terrorism. Winning the "war on terrorism" was the first priority, and accomplishing it would require commitment to two other top priorities: enhancing joint warfighting and "transforming" the forces for the future. The document described a security environment characterized by a wide range of adversaries (from traditional military forces to nonstate organizations and rogue states). The diversity of anticipated threats meant that the armed forces needed to be prepared to operate in a more complex and distributed battlespace. The 2004 NMS also noted that technological diffusion was changing the nature of that battlespace, as potentially dangerous dual-use civilian technologies were increasingly available to adversaries. To operate in this environment, the NMS called on the armed forces to continue developing capabilities to remain agile, adaptable, integrated, and expeditionary.

[6]   U.S. Joint Chiefs of Staff, *The National Military Strategy of the United States of America*, Washington, D.C., 2004. Since this writing, an updated NMS was issued (February 8, 2011).

Quadrennial Defense Review (2010)

The 2010 *Quadrennial Defense Review Report* reflects a vision of a complex, upcoming range of dynamic international security challenges and a vision for orienting DoD resources to meet them.[7] Globally, the QDR describes demographic, technological, economic, and environmental trends expected to add complexity to international relations. It points to China and India as rising international players, indicative of the extent to which the United States can no longer go it alone in its effort to sustain international peace and stability. It also notes that rapid technological change is altering both the state of global connectivity and the conduct of war.

The report prioritizes success both in current operations and across a range of disparate and dynamic future threats. In the near term, DoD is to prioritize prevailing in operations in Afghanistan and Iraq and supporting Afghan and Pakistani leadership in successfully disrupting, dismantling, and defeating al Qaeda. Beyond current operations, the QDR tasks DoD with focusing on preventing and deterring future conflicts through the projection of balanced military force as well as such means as diplomacy, development, and intelligence. Should deterrence fail, DoD's third priority is to prepare to defeat adversaries and succeed across a range of contingencies.

The balanced force that would meet these strategic objectives would rely in many ways on sound intelligence capabilities. The classes of operations for which DoD needs to enhance capabilities are COIN, stability, and counterterrorism. To this end, the QDR report calls for the expansion of systems to support ISR activities, regional expertise, and strategic communication. To help build the security capacity of partner states, the report specifies the need for enhanced linguistic, regional, and cultural expertise. Intelligence would also be key to deterring and defeating potentially hostile nation-states, which are difficult to penetrate by other means. Specifically, robust ISR capabilities, space-based systems, and sensors would be key assets against nation-states. Many of these capabilities rely more broadly on effective operations in

---

[7] U.S. Department of Defense, *Quadrennial Defense Review Report*, Washington, D.C., February 2010a.

cyberspace and thus greater centralized command of cyber operations and enhanced coordination with other agencies and governments.

Quadrennial Intelligence Community Review (January 2009)

The DNI released the unclassified version of the *Quadrennial Intelligence Community Review* in January 2009.[8] This document presents the DNI's perspective on alternative futures, future missions for the IC, and operating principles and required capabilities to fulfill those missions. Like other planning guidance, it describes a national security environment characterized by unpredictable and complex threats. The IC needs to respond rapidly and employ more innovative analytical techniques and collection means. Achieving this requires the community to organize around missions rather than collection stovepipes and to exploit technical networks to integrate activities. The report considers four alternative futures, two of which prioritize the role of state actors and two in which nonstate actors play the dominant role: (1) China/Russia/India/Iran-centered bloc that sets the pace for innovative technologies to challenge U.S. global predominance; (2) precarious balance of power resulting from states locked in multipolar competition jockeying for resources; (3) power shifting to nonstate actors, such as corporations or megacities, allowing global ills to spiral out of control; (4) identity-based groups supplant the authority of nation-states, competing with one another for influence in a chaotic political environment.

Report on Progress Toward Security and Stability in Afghanistan and United States Plan for Sustaining the Afghanistan National Security Forces (April 2010)

This report is the product of a congressional mandate for the administration to report to Congress every 180 days on stability and strategy

---

[8] Office of the Director of National Intelligence, *Quadrennial Intelligence Community Review: Alternative Futures the IC Could Face*, Washington, D.C., January 2009a.

in Afghanistan.[9] It represents the coordinated efforts of the Secretary of Defense, the Secretary of State, the DNI, the U.S. Attorney General, the administrator of the Drug Enforcement Agency, the administrator of the U.S. Agency for International Development, the Secretary of Agriculture, and the Secretary of the Treasury.

Much less abstract than other national planning documentation, the report describes the visceral and immediate strategic environment of combat operations, including the threat of al Qaeda in Afghanistan and Pakistan and the destabilizing effect of the Taliban. On the ground, the strategy of U.S. forces in the April 2010 report is COIN, with an emphasis on population security, counterterrorism operations, and efforts to build Afghan security forces. The nature of the work is manpower-intensive, and the strategy calls for an additional 30,000 U.S. troops. To facilitate operations in this environment, the report calls for a wide range of intelligence activities: ongoing intelligence efforts to counter IED attacks, improve situational awareness, and facilitate raids, airstrikes, stability operations, and humanitarian efforts. Additionally, some intelligence sharing has been institutionalized through the Tripartite Joint Intelligence Operation Center, which fosters coordination and cooperation among the International Security Assistance Force and Afghan and Pakistani forces.

Marine Corps Planning Documents

A Cooperative Strategy for a 21st Century Seapower (2007)

A Cooperative Strategy for a 21st Century Seapower was released by the Chief of Naval Operations, Commandant of the Coast Guard, and Commandant of the Marine Corps in October 2007.[10] It was guided by the NSS, NDS, NMS, and the National Strategy for Maritime

---

[9] U.S. Department of Defense, *Report on Progress Toward Security and Stability in Afghanistan and United States Plan for Sustaining the Afghan National Security Forces*, Washington, D.C., April 2010b.

[10] See U.S. Marine Corps, U.S. Navy, and U.S. Coast Guard, *A Cooperative Strategy for 21st Century Seapower*, Washington, D.C., October 2007.

Security and was intended to highlight a number of potential future challenges that the Navy will face. First, globalization and continued growth could create increased competition for resources between nations. This competition might encourage nations to exert wider claims of sovereignty over oceans, waterways, and natural resources. In addition, the document notes that globalization is shaping the conduct of conflicts, increasingly characterized by a hybrid blend of traditional and irregular tactics. Second, while the expansion of new technologies offers opportunities, it also threatens to become a source of competition and conflict for access and natural resources. The document predicts that asymmetric use of technology could present a wide range of threats to the United States and its allies. These conditions, combined with the effects of population growth and climate change, create an uncertain future.

According to the report, the Maritime Strategic Concept asserts that U.S. maritime forces will be characterized as "regionally concentrated, forward-deployed task forces with the combat power to limit regional conflict and deter major power war."[11] To protect U.S. interests, combat power must be continuously postured in the western Pacific, Arabian Gulf, and Indian Ocean. In addition, the maritime forces must be tailored to meet the requirements of each geographic region. To implement this strategy, the Navy, Coast Guard, and USMC must expand the core capabilities of U.S. seapower. Maritime forces will be forward deployed and must demonstrate flexibility and adaptability to meet future challenges. The report also noted future effectiveness will require an increased commitment to advance maritime domain awareness and expanded ISR capability and capacity.

Marine Corps Vision and Strategy 2025
The *Marine Corps Vision and Strategy 2025*, released in 2008, serves as the principal strategic planning document for Marine Corps roles, functions, and composition.[12] It derived its vision for the USMC from national and DoD guidance, such as the NSS, NDS, NMS, and QDR.

---

[11] See U.S. Marine Corps, U.S. Navy, and U.S. Coast Guard, 2007, p. 8.

[12] Commandant of the Marine Corps, 2008.

The current report outlines a plan for operating as the nation's premier expeditionary force in an inherently uncertain strategic environment. It also nests Marine Corps strategy within the broad global context. With regard to demographics, the document projects global population growth, with an increase in urban populations in Asia and in Africa. In an economic context, it predicts that globalization would continue, increasing interactions between societies and increasing demand for resources.

Within this broad context, *Marine Corps Vision and Strategy 2025* projects that future threats would most likely be "hybrid" in character, blurring once-distinct challenges, such as conventional war, irregular challenges, terrorism, and criminality. Hybrid challenges, it notes, could arise from adversaries ranging from states to nonstate actors, proxy forces, or armed groups. It anticipates that adversaries will blend different approaches and integrate various weapons, tactics, and technologies. Meeting these challenges would be further complicated by the increased complexity of future operational environments, likely to be denser, more populated, and more urban than in the past.

Operating in such a strategic context would require the USMC to be innovative in the organization of its forces and resources. Its operational effectiveness is founded on the integrated MAGTF. The report calls for qualitative changes to the MAGTF operating force structure to enhance small-unit training and situational awareness and to reduce gaps in tactical mobility and assault support. It identifies a key element of such improvements as the broader application of unmanned systems and integrated ISR activities. It also calls for proper equipping, with intelligence systems down to the CE, allowing marines to understand the specific environment, detect and locate threats, and provide useful and timely intelligence at all levels. The *Marine Corps Vision and Strategy 2025* puts forth a plan to invest in integrating C2 and ISR capabilities all the way down to the squad level.

Marine Corps Intelligence Surveillance and Reconnaissance Enterprise Roadmap (2010)

DIRINT, BGen. Vincent R. Stewart, described the *MCISR-E Roadmap* as a plan for realizing the vision put forth in the *Marine Corps*

Vision and Strategy 2025 and the Service Campaign Plan.[13] Those planning documents described a future defined by a complex *hybrid* threat environment. Future conflicts would pit U.S. forces against adversaries employing both primitive and sophisticated technologies, engaging in both irregular and conventional tactics, and operating in complex physical and political environments. This result would be a USMC intelligence organization that had to be "continuously operational"—prepared to meet a range of challenges, even in times of peace.[14] The *MCISR-E Roadmap* also anticipates a changing fiscal environment. With the drawdown of combat operations in Iraq and Afghanistan, the document assumed that the ends of the USMC intelligence would not be achieved through increases in end strength or in budget levels.

The roadmap introduced the concept of the "enterprise," which it put forth as a means of optimizing the operation of USMC ISR capabilities. The enterprise consists of the collective personnel, equipment, and organizations in both the supporting establishment and the operating forces with ISR responsibilities. It describes how the enterprise can be built into an entity capable of operating quickly and flexibly in a complex environment. Other priorities include growing a professional intelligence workforce equipped with appropriate skills and expertise and conducting the analyses necessary to anticipate threats and identify key emerging technologies.

Building the enterprise would require integration and coordination at all levels, as well as "synergistic integration" of all ISR elements both in the MAGTF and in the supporting establishment (including the I-Dept and MCIA). Furthermore, the roadmap specifies a need to reach outside the organization to the IC and nontraditional partners. Thus, integration would be achieved by means of a shared vision and shared resources. The result would be a more networked organization, with free flows of information, integrated data management, and common materiel solutions.

---

[13] See U.S. Marine Corps Intelligence Department, 2010.

[14] U.S. Marine Corps Intelligence Department, 2010, p. 5.

Details of Alternative Structure Assessments

In addition to our holistic assessment of the fit of our proposed structural alternatives according to organizational design criteria, we also conducted a more formal assessment. The details of this effort are presented here; the results are presented in Chapter Seven.

To assess the alternatives at each level, we first identified the appropriate assessment criteria. These criteria were drawn from identified end states specific to the organizational level and concerns raised (drawn from the issues identified in Chapters Six and Eight) about that organizational level. We then arrived at five categories based on the discussion in Chapter Four: goals, strategy, resources and authority, environment, and structure.

We assessed each alternative (including an as-is/base case) against each specific end state or concern. Assessments were numerical scores, 1–3, where 1 indicated improvement needed, 2 indicated adequate, and 3 indicated good. These scores were highly subjective—informed by our experience and research, to be sure—but subjective nonetheless.

In this appendix, we present the criteria and the assessments for each alternative at each organizational level.

Assessment of Intelligence Department Alternatives

Table F.1 presents the criteria on which we assessed I-Dept, the end state, and the concerns (the columns), categorized by the five organizational dimensions of interest.

Table F.1
End States and Concerns for I-Dept

Category	End State	Concerns
Goals	Input to USMC and IC policy and resource processes Premium on efficient management processes	No long-term, strategic focus No connection to operations (CE, MEF)
Strategy	Able to exploit process to the benefit of USMC and USMC intelligence	Stovepiped functional analysis Organization opaque to outsiders so difficult to engage
Resources and authority	Experienced functional staff Centralized decisions	Experienced but low grade structure Vacancies in management Low numbers, little policy experience DIRINT has limited authority
Environment	Complex and unpredictable	Will remain the same
Structure	Functionally aligned hierarchy	Sections misnamed

Table F.2 presents our assessment of the three alternatives (including the as-is/base case) for I-Dept. As noted, each criterion is scored 1–3 (poor to good) for each alternative. Summing the scores for each alternative (bottom row in Table F.2) confirms realignment as the best, most reliable approach. The shaded rows are the assessments of the end states in Table F.1, and the unshaded rows are the assessments for the concerns.

Assessment of MCIA Alternatives

Table F.3 reproduces Table F.2 for MCIA, showing both end states and concerns specific to MCIA, categorized by organizational dimensions.

Table F.4 presents our assessment of the three alternatives (including the as-is/base case) for MCIA. Summing the scores for each alternative (bottom row in Table F.4) confirms a matrix organization as the best approach.

Table F.2
Assessment of Alternatives for I-Dept

I-Dept Assessment	As Is	Realign	Matrix
Goals			
Efficiently engage in USMC, DoD, IC process	2	3	1
Capability for long-term, strategic focus	1	3	Depends
Facilitate input from operations (CE, MEF)	1	2	Depends
Strategy			
Exploit process	2	3	3
Stovepiped functional analysis	1	3	3
Opaque to outsiders	1	3	Depends
Resources and authority			
Experienced functional staff expertise	2	2	2
Centralized decisions	2	2	Depends
Experienced but low grade structure	2	2	2
Vacancies in management	1	1	2
Low numbers, policy experience	2	2	1
DIRINT has limited authority	1	1	1
Environment			
Complex and predictable	2	2	2
Structure			
Functionally aligned hierarchy	2	2	1
Names of sections wrong	1	3	Depends
Summary	23	34	23–33

Table F.3
End States and Concerns for MCIA

End State	Concerns
Goals	
Produce intelligence products for range of customers (up and down)	Mission priorities not clear
Support intelligence DOTMLPF	Customer priorities not clear
Lead for cultural intelligence	Website not customer-friendly
Fixed site for USMC integration	Lack of 24/7 watch cycle
Premium on efficient and innovative production	Not effectively used
Strategy	
Produce and innovate within functions	Need more functional integration focused on customer or tasks
Resources and authority	
Substantial staff assets (civilian and military)	Serve multiple masters, especially DIRINT
Experienced, functional experts	Complex coordination processes
Organizations have clear hierarchies	Resources assigned to priorities
Decentralized production	NA
Environment	
Complex, relatively predictable	Environment may become less predictable
Structure	
Hierarchical, many subordinate commands	Excessive bureaucracy

Table F.4
Assessment of Alternatives for MCIA

End State	As Is	Divisional by Customer	Customer/ Functional Matrix
Goals			
Produce intelligence products for range of customers (up and down)	2	3	3
Support intelligence DOTMLPF	2	3	3
Lead for cultural intelligence	3	2	3
Fixed site for USMC intelligence	1	2	3
Premium on efficient and innovative production	1	2	3
Mission priorities not clear	2	3	3
Customer priorities not clear	1	3	3
Website not customer-friendly	1	1	2
Lack of 24/7 watch cycle	1	3	3
Not effectively used	1	3	3
Strategy			
Produce and innovate within functions	2	1	3
More functional integration focused on customer or tasks	1	2	3
Resources and authority			
Substantial staff assets (civilian and military)	3	3	3
Experienced, functional experts	3	3	3
Organizations have clear hierarchies	2	2	1
Decentralized production	2	3	3
Serve multiple masters, especially I-Dept	1	3	3
Complex coordination processes	1	2	1
Make sure resources assigned to priorities	1	2	3

Table F.4—Continued

End State	As Is	Divisional by Customer	Customer/ Functional Matrix
Environment			
Complex, relatively predictable	2	3	2
May become less predictable	1	2	3
Structure			
Hierarchical, many subordinate commands	2	3	3
Excessive bureaucracy	1	1	3
Summary	37	55	63

Assessment at the MEF Level

Table F.5 reproduces Table F.1 for the MEF intelligence structures, specifically the intelligence and radio battalions, showing both end states and concerns specific to the MEFs, categorized by organizational dimensions.

Table F.6 presents our assessment of the three alternatives (including the as-is/base case) for the MEF. Summing the scores for each alternative (bottom row in Table F.6) confirms that a matrix organization comprising the intelligence and the radio battalion structure as the best approach.

Table F.5
End States and Concerns for MEF Intelligence

End State	Concerns
Goals	
Produce intelligence for range of customers (up and down)	CE finds support lacking—not relevant, not timely, adds to CE requirements
	In garrison, intelligence battalion does minimal intelligence work
	Products not sufficiently integrated across functions
Strategy	
Exploit and innovate across functions	Competing missions, priorities—up wins
	Does not understand CE customer
Resources and authority	
Many assets assigned	In garrison, not enough classified network access
Resource allocation depends on personalities	
Range of experience, expertise	Authority conflated between intelligence battalion commander and MEF G-2
	Trains as intelligence battalion but does not deploy as a battalion
	Intelligence professionals used for administrative, management, oversight, and command tasks
Environment	
Complex and unpredictable	Likely to continue for deployed forces
	Preparing for breadth will become more important
Structure	
Functional and divisional	Need for command billets leads to copying other USMC occupational field structures

Table F.6
Assessment of Alternatives for MEF

End State	As Is	Matrix: Intelligence Battalion	Matrix: Intelligence and Radio Battalion
Goals			
Produce intelligence for range of customers (up and down)	2	3	3
CE finds support lacking	1	2	3
In garrison, intelligence battalion does minimal intelligence work	1	2	3
Products not sufficiently integrated	1	2	3
Strategy			
Exploit and innovate across functions	1	2	3
Competing missions, priorities—up wins	1	3	3
Does not understand CE customer	2	3	3
Resources and authority			
Many assets assigned	3	3	3
Resource allocation depends on personalities	1	1	1
Range of experience, expertise	3	2	3
In garrison, not enough classified network access	1	2	1
Authority conflated between intelligence battalion commander and MEF G-2	1	1	1
Trains as intelligence battalion but does not deploy as a battalion	1	2	3
Intelligence professionals used for administrative, management, oversight, and command tasks	1	1	1
Environment			
Complex and unpredictable	2	2	3
Likely to continue for deployed forces	2	2	3

Table F.6—Continued

End State	As Is	Matrix: Intelligence Battalion	Matrix: Intelligence and Radio Battalion
Preparing for breadth will become more important	1	2	3
Structure			
Functional and divisional	1	2	3
Need for command billets leads to copying other USMC occupational field structures	2	1	1
Summary	28	38	47

Assessment of Combat Element Intelligence

Table F.7 reproduces Table F.1 for the combat element (GCE, ACE, and LCE) intelligence structures, showing both end states and concerns specific to the combat elements, categorized by organizational dimension.

Table F.8 presents our assessment of the three alternatives considered for the combat elements. Summing the scores for each alternative (bottom row in Table F.8) confirms that a matrix organization with CLIC is the best approach.

Table F.7
End States and Concerns for the Combat Elements

Objectives	Concerns
Goal	
Produce actionable intelligence for owning operating force, others as tasked	Cannot do what S3 wants Collaboration can help a lot
Strategy	
Innovate own collection and analysis	Innovation hampered by S3's (and others') lack of experience using intelligence
Exploit others' collection and analysis	No connectivity, access is personality-based Have not captured and transferred good innovations Do not habituate relationships in training
Resources and authority	
Personnel have mixed, often limited expertise	Inexperienced personnel not helpful to unit nor intelligence effort generally
Sometimes augmented by detachments	More tasking than support from above Training not sufficient
Authority over intelligence subject to S3 priorities	
Environment	
Complex and unpredictable	Likely to continue in COIN
Narrowly focused mission set	Will get broader post-OEF
Structure	
Dual hierarchy of function and division	Causes friction setting priorities, integrating intelligence with customer
Matrix-like within unit	Functions do not train as integrated team

Table F.8
Assessment of Alternatives for the Combat Elements

End State	As Is	S2 in Charge	Matrix S2 and CLIC
Goals			
Produce actionable intelligence for owning operating force, others as tasked	2	3	3
Cannot do what S3 wants	1	2	3
Collaboration can help a lot	1	2	3
Strategy			
Innovate own collection and analysis	2	1	3
Exploit others' collection and analysis	1	2	3
Innovation hampered by S3's (and others') lack of experience using intelligence	2	2	3
No connectivity, access is personality-based	1	2	3
Have not captured and transferred good innovations	1	1	2
Do not habituate relationships in training	1	1	2
Resources and authority			
Personnel have mixed, often limited expertise	1	2	3
Sometimes augmented by detachments	1	1	2
Authority over intelligence subject to S3 priorities	2	2	3
Inexperienced personnel not helpful to unit nor intelligence generally	1	2	1
More tasking than support from above	1	2	2
Training not sufficient	1	2	3
Environment			
Complex and unpredictable	2	2	2
Narrowly focused mission set	2	2	3

Table F.8—Continued

End State	As Is	S2 in Charge	Matrix S2 and CLIC
Environment (continued)			
Likely to continue in COIN	2	2	2
Will get broader post-OEF	1	1	2
Structure			
Dual hierarchy of function and division	2	3	1
Matrix-like within unit	2	3	3
Causes friction setting priorities, integrating intelligence with customer	2	3	2
Functions do not train as integrated team	1	2	3
Summary	33	45	57

Bibliography

All Marines Memo 008/07, "Marine Corps End Strength Increase," February 7, 2007.

All Marines Memo 021/00, "Establishment of Intelligence Department (Code I) at HQMC," 2000.

All Marines Memo 100/95, "Program to Improve Marine Corps Intelligence," March 24, 1995.

Barron, F. Hutton, and Bruce E. Barrett, "Decision Quality Using Ranked Attribute Weights," *Management Science*, Vol. 42, No. 11, November 1996, pp. 1515–1523.

Best, Richard A., Jr., *Proposals for Intelligence Reorganization: 1949–2004*, Washington, D.C.: Congressional Research Service, RL32500, September 24, 2004a.

———, *Intelligence Community Reorganization: Potential Effects on DoD Intelligence Agencies*, Washington, D.C.: Congressional Research Service, RL32515, December 21, 2004b.

———, *Intelligence, Surveillance, and Reconnaissance (ISR) Programs: Issues for Congress*, Washington D.C.: Congressional Research Service, RL32508, February 22, 2005.

———, *Intelligence Issues for Congress*, Washington, D.C.: Congressional Research Service, September 18, 2009.

Boren, David L., "The Intelligence Community: How Crucial?" *Foreign Affairs*, Summer 1992.

Buikema, R. J., *Integration of Intelligence into Professional Military Education*, thesis, Quantico, Va.: U.S. Marine Corps Command and Staff College, April 18, 1996.

Burton, Richard M., Gerardine DeSanctis, and Børge Obel, *Organizational Design: A Step-By-Step Approach*, Cambridge, UK: Cambridge University Press, 2006.

C4I Staff, Headquarters, U.S. Marine Corps, "The Future of Marine Corps Intelligence," *Marine Corps Gazette*, Vol. 78, No. 4, April 1995, pp. 26–29.

Chizek, Judy G., *Military Transformation: Intelligence, Surveillance and Reconnaissance*, Washington, D.C.: Congressional Research Service, RL31425, January 17, 2003.

Coia, Raymond E., *A Critical Analysis of the I MEF Intelligence Performance in the 1991 Persian Gulf War*, Quantico, Va.: U.S. Marine Corps Command and Staff College, May 22, 1995.

Collins, Matthew, "Beyond the Van Riper Plan: How Are We Growing Intelligence Officers," *Marine Corps Gazette*, Vol. 90, No. 10, October 2006.

Colvard, C. E., "Unfortunately, We Fought Like We Trained," *Marine Corps Gazette*, Vol. 75, No. 9, September 1991.

Commandant of the Marine Corps, *Marine Corps Vision and Strategy 2025*, Arlington, Va.: Office of Naval Research, 2008. As of March 17, 2011: http://www.onr.navy.mil/~/media/Files/About%20ONR/usmc_vision_strategy_2025_0809.ashx

Conway, James T., "A Concept for Enhanced Company Operations," *Marine Corps Gazette*, Vol. 92, No. 12, December 2008.

Cumming, Alfred, *The Position of Director of National Intelligence: Issues for Congress*, Washington, D.C.: Congressional Research Service, RL32506, August 12, 2004.

Daggett, Stephen, *The U.S. Intelligence Budget: A Basic Overview*, Washington, D.C.: Congressional Research Service, RL21945, September 24, 2004.

Davidoski, Joseph, "More Than Mapmakers," *Marine Corps Gazette*, Vol. 94, No. 9, September 2010.

Decker, Michael H., "Assessing the Intelligence Effort," *Marine Corps Gazette*, Vol. 75, No. 9, September 1991.

Defense Science Board, *Transformation Study Report: Transforming Military Operational Capabilities*, Washington, D.C., April 27, 2001.

DeLeal, William E., "Finding a Needle in a Stack of Needles," *Marine Corps Gazette*, Vol. 93, No. 1, January 2009.

Deputy Commandant for Combat Development and Integration, U.S. Marine Corps, *Marine Corps Operating Concepts*, 3rd ed., Quantico, Va., June 2010. As of March 17, 2011: http://www.quantico.usmc.mil/uploads/files/MOC%20July%2013%20update%202010_Final.pdf

Dinsmore, Jeffrey S., "Intelligence Support to Counterinsurgency Operations: The Search for Fused, Coherent Intelligence to Support the Commander," *Marine Corps Gazette*, Vol. 91, No. 7, July 2007, pp. 13–16.

Druckman, Daniel, Jerome E. Singer, and Harold Van Cott, eds., *Enhancing Organizational Performance*, Washington D.C.: National Academies Press, 1997.

Flynn, Michael T., Matt Pottinger, and Paul D. Batchelor, *Fixing Intel: A Blueprint for Making Intelligence Relevant in Afghanistan*, Washington, D.C.: Center for a New American Security, January 4, 2010a.

———, "Fixing Intel in Afghanistan," *Marine Corps Gazette*, Vol. 94, No. 4, April 2010b.

Foley, Michael P., "Facilitating Intelligence at the Point of Action," *Marine Corps Gazette*, Vol. 94, No. 3, March 2010, pp. 16–19.

Galbraith, Jay R., *Designing Matrix Organizations That Actually Work: How IBM, Procter & Gamble and Others Design for Success*, San Francisco, Calif.: Jossey-Bass, 2009.

Gates, Robert M., Secretary of Defense, "SECDEF Statement," Washington, D.C., August 9, 2010. As of March 15, 2011:
https://dap.dau.mil/policy/Documents/Policy/Efficiencies%20Statement%20As%20Prepared.pdf

Goulding, Vincent J., Jr., "Enhanced Company Operations," *Marine Corps Gazette*, Vol. 92, No. 8, August 2008.

Graham, Edward P., "Company-Level Intelligence Cell," *Marine Corps Gazette*, Vol. 94, No. 3, March 2010.

Grogan, M. S., S. Lima, J. Terando, and G. A. Winterstein, *The Surveillance, Reconnaissance, Intelligence Group Concept and Organization*, Quantico, Va.: Communication Officers School, March 23, 1992.

Headquarters, U.S. Department of the Army, *Operations*, Washington D.C., Field Manual 3-0, February 27, 2008.

Headquarters, U.S. Department of the Army, and Headquarters, U.S. Marine Corps, *Operational Terms and Graphics*, Washington, D.C., Field Manual 101-5-1/Marine Corps Reference Publication 5-2A, September 30, 1997.

Headquarters, U.S. Marine Corps, *Organization of Marine Corps Forces*, Washington, D.C., Marine Corps Reference Publication 5-12D, October 13, 1998.

———, *Intelligence Operations*, Washington, D.C., Marine Corps Warfighting Publication 2-1, September 10, 2003.

———, *Marine Corps 2005–2015: ISR Roadmap*, draft, October 2, 2006.

———, *Reshaping America's Expeditionary Force in Readiness: Report of the 2010 Marine Corps Force Structure Review Group*, Washington, D.C., March 14, 2011.

Higgins, James L., Michelle L. Trusso, and Alfred B. Connable, "Marine Corps Intelligence," *Marine Corps Gazette*, Vol. 89, No. 12, December 2005.

Hisey, Braden W., "Producing a More Practical Tactical Intelligence Officer," *Marine Corps Gazette*, Vol. 82, No. 12, December 1998, pp. 13–15.

Hoffman, F. G., "The Corps' Expansion," *Marine Corps Gazette*, Vol. 91, No. 6, June 2007.

Huddleston, Craig, "Commentary on Desert Shield," *Marine Corps Gazette*, Vol. 75, No. 6, June 1991, pp. 32–33.

Johnston, John W., "A Marine Corps Intelligence/Signals Intelligence/Electronic Warfare Perspective," *Marine Corps Gazette*, Vol. 79, No. 1, January 1995, pp. 17–18.

Joint Defense Science Board Intelligence Science Board Task Force, *Integrating Sensor-Collected Intelligence*, Washington, D.C., November 2008.

Lamothe, Dan, "Kent Takes on Drawdown Rumors, PFT Fairness," *Marine Corps Times*, October 1, 2010. As of October 5, 2010:
http://www.marinecorpstimes.com/news/2010/10/marine-corps-sgt-maj-carlton-kent-afghanistan-drawdown-100110/

Lawler, Edward E. *From the Ground Up: Six Principles for Building the New Logic Corporation*, San Francisco, Calif.: Jossey-Bass, 1996.

Liebl, Vernie R., "The Intelligence Plan: An Update," *Marine Corps Gazette*, Vol. 85, No. 1, January 2001.

Lively, James W., "Cultural Education," *Marine Corps Gazette*, Vol. 91, No. 4, April 2007.

Livingston, Robert W., "Marine Corps Intelligence Activity—Excellence in Expeditionary Intelligence," *Marine Corps Gazette*, Vol. 79, No. 4, April 1995.

Lowenthal, Mark M., *Intelligence: From Secrets to Policy*, Washington, D.C.: Congressional Quarterly Press, 2006.

Mann, Morgan G., and Michael Driscoll, "Thoughts Regarding the Company-Level Intelligence Cell," *Marine Corps Gazette*, Vol. 93, No. 6, June 2009.

Marine Administrative Message 079/01, "Command Activation," February 2001.

Marine Corps Order 3900.15B, "Marine Corps Expeditionary Force Development System (EFDS)," March 10, 2008.

McMains, James W., "The Marine Corps Robotics Revolution," *Marine Corps Gazette*, Vol. 88, No. 1, January 2004.

Michael, E. Ennis, "The Future of Intelligence," *Marine Corps Gazette*, Vol. 83, No. 10, October 1999.

Nicola, Thomas J., *9/11 Commission Recommendations: Intelligence Budget*, Washington, D.C.: Congressional Research Service, September 27, 2004.

O'Rourke, Ronald, *Defense Transformation: Background and Oversight Issues for Congress*, Washington, D.C.: Congressional Research Service, RL32238, November 9, 2006.

Office of the Director of Intelligence, U.S. Marine Corps, *ISR Roadmap*, Washington D.C., October 2, 2006.

Office of the Director of National Intelligence, *Vision 2015: A Globally Networked and Integrated Intelligence Enterprise*, Washington, D.C., 2008. As of March 17, 2011:
http://www.dni.gov/Vision_2015.pdf

———, *Quadrennial Intelligence Community Review: Alternative Futures the IC Could Face*, Washington, D.C., January 2009a.

———, *National Intelligence Strategy*, Washington, D.C., August 2009b. As of March 17, 2011:
http://www.dni.gov/reports/2009_NIS.pdf

Office of the President of the United States, *National Security Strategy*, Washington, D.C., May 2010. As of March 17, 2011:
http://www.whitehouse.gov/sites/default/files/rss_viewer/national_security_strategy.pdf

Reiley, Matthew A., *Transforming USMC Intelligence to Address Irregular Warfare*, thesis, Quantico, Va.: U.S. Marine Corps Command and Staff College, 2008.

Rumsfeld, Donald, "Transforming the Military," *Foreign Affairs*, Vol. 81, No. 3, May–June 2002.

Smist, Frank J., Jr., *Congress Oversees the United States Intelligence Community, 1947–1994*, Knoxville, Tenn.: University of Tennessee Press, 1994.

Takle, Jeffrey N., "The Intelligence Plan: A Three-Legged Chair?" *Marine Corps Gazette*, Vol. 86, No. 2, February 2002, pp. 28–30.

Taylor, Daniel P., "Eyes in the Sky: Unmanned Aerial Vehicles Expand Marine Corp ISR Capabilities," *Seapower*, April 2010.

Treverton, Gregory F., *Reshaping National Intelligence for an Age of Information*, New York: Cambridge University Press, 2001.

U.S. Army Intelligence Center and Ft. Huachuca, *Combat Commander's Handbook on Intelligence*, Ft. Huachuca, Ariz., Special Text No. 2-50.4 (Field Manual 34-8), September 2001.

U.S. Army Training and Doctrine Command, *The U.S. Army Functional Concept for Intelligence 2016–2028*, Fort Monroe, Va., Pamphlet 525-2-1, October 13, 2010.

U.S Department of Defense, *Quadrennial Defense Review Report*, Washington, D.C., September 30, 2001.

————, *National Defense Strategy*, Washington, D.C., June 2008. As of March 17, 2011:
http://www.defense.gov/news/2008%20national%20defense%20strategy.pdf

————, *Quadrennial Defense Review Report*, Washington, D.C., February 2010a. As of March 17, 2011:
http://www.defense.gov/qdr/images/QDR_as_of_12Feb10_1000.pdf

————, *Report on Progress Toward Security and Stability in Afghanistan and United States Plan for Sustaining the Afghan National Security Forces*, Washington, D.C., April 2010b. As of March 17, 2011:
http://www.defense.gov/pubs/pdfs/Report_Final_SecDef_04_26_10.pdf

————, *Measuring Stability and Security in Iraq*, quarterly report to Congress, Washington, D.C., June 2010c. As of March 17, 2011:
http://www.defense.gov/home/features/iraq_reports

U.S. House of Representatives, "The Current and Future State of Intelligence," hearing before the Permanent Select Committee on Intelligence, Washington, D.C., February 24, 1994.

————, "2010 Posture of the United States Marine Corps," hearing before the Committee on Armed Services, Washington, D.C., February 24, 2010.

U.S. Joint Chiefs of Staff, *The National Military Strategy of the United States of America*, Washington, D.C., 2004. As of March 17, 2011:
http://www.defense.gov/news/mar2005/d20050318nms.pdf

U.S. Marine Corps, U.S. Navy, and U.S. Coast Guard, *A Cooperative Strategy for 21st Century Seapower*, Washington, D.C., October 2007. As of March 17, 2011:
http://www.navy.mil/maritime/Maritimestrategy.pdf

U.S. Marine Corps Center for Lessons Learned, *Counterintelligence/Human Intelligence Exploitation Operations: Quick Look Report*, Washington, D.C., 2008.

U.S. Marine Corps Intelligence Department, "'202K' Build Out for Marine Corps Intelligence," Washington, D.C., undated.

————, *The Marine Corps Intelligence, Surveillance, and Reconnaissance Enterprise (MCISR-E) Roadmap*, Washington D.C., April 28, 2010.

U.S. Senate, "Marine Corps Intelligence Programs and Lessons Learned in Recent Military Operations," hearing before the Committee on Armed Services, Subcommittee on Strategic Forces, Washington, D.C., April 7, 2004.

Van Riper, Paul K., "Observations During Operation Desert Storm," *Marine Corps Gazette*, Vol. 75, No. 6, June 1991, pp. 55–61.

Vernie, R. Liebl, "The Intelligence Plan: An Update," *Marine Corps Gazette*, Vol. 85, No. 1, January 2001.

Wear, John M., "Educating Intelligence Specialists," *Marine Corps Gazette*, Vol. 93, No. 5, May 2009.

Wright, Donald P., and Timothy R. Reese, *On Point II, Transition to the New Campaign: The United States Army in Operation Iraqi Freedom, May 2003– January 2005*, Ft. Leavenworth, Kan.: Combat Studies Institute Press, June 2008.